AF806815

Scholarly Virtues in Nineteenth-Century Sciences and Humanities

Christiaan Engberts

Scholarly Virtues in Nineteenth-Century Sciences and Humanities

Loyalty and Independence Entangled

palgrave
macmillan

Christiaan Engberts
Utrecht University
Utrecht, The Netherlands

ISBN 978-3-030-84565-0 ISBN 978-3-030-84566-7 (eBook)
https://doi.org/10.1007/978-3-030-84566-7

© The Editor(s) (if applicable) and The Author(s), under exclusive licence to Springer Nature Switzerland AG 2022
This work is subject to copyright. All rights are solely and exclusively licensed by the Publisher, whether the whole or part of the material is concerned, specifically the rights of translation, reprinting, reuse of illustrations, recitation, broadcasting, reproduction on microfilms or in any other physical way, and transmission or information storage and retrieval, electronic adaptation, computer software, or by similar or dissimilar methodology now known or hereafter developed.
The use of general descriptive names, registered names, trademarks, service marks, etc. in this publication does not imply, even in the absence of a specific statement, that such names are exempt from the relevant protective laws and regulations and therefore free for general use.
The publisher, the authors and the editors are safe to assume that the advice and information in this book are believed to be true and accurate at the date of publication. Neither the publisher nor the authors or the editors give a warranty, expressed or implied, with respect to the material contained herein or for any errors or omissions that may have been made. The publisher remains neutral with regard to jurisdictional claims in published maps and institutional affiliations.

This Palgrave Macmillan imprint is published by the registered company Springer Nature Switzerland AG.
The registered company address is: Gewerbestrasse 11, 6330 Cham, Switzerland

Acknowledgements

The research for this book has been made possible by the Netherlands Organization for Scientific Research (NWO) who funded the research *The Scholarly Self: Character, Habit and Virtue in the Humanities, 1860–1930* between 2013 and 2018 at Leiden University. Of course, my research would not have been the same if it were not for the input of the project's leader, Prof. Herman Paul, and my closest colleagues in the project, Katharina Manteufel and Dr. Léjon Saarloos.

The research carried out for this book has also been made possible by two extended stays in Germany. I would like to thank Prof. Matthias Middell and Dr. Martina Keilbach at the Global and European Studies Institute at the University of Leipzig for hosting me in the autumn of 2015. I would like to thank Prof. Arnd Bauerkämper at the Friedrich-Meinecke-Institut of the Freie Universität Berlin for hosting me during the autumn of 2016.

In addition, I would like to thank the helpful staff of the different libraries and archives, without whom my research would not have been possible. I am particularly grateful to the staff at Leiden University Library, the *Universitätsarchiv* of the Martin-Luther-University in Halle, the *Universitätsarchiv* and the *Universitätsbibliothek* of the University of Leipzig, and the *Geheimes Staatsarchiv Preußischer Kulturbesitz*, the *Staatsbibliothek*, and the *Robert Koch-Institut* in Berlin.

Contents

CHAPTER 1

Introduction

The (in)gratitude of Hugo Münsterberg

Even though the Leipzig philosopher Wilhelm Wundt was not known for his charismatic presence and attractive presentation, one of his 1883 lectures made a profound impression on at least one ambitious member of his audience. The young medical student Hugo Münsterberg was so captivated that he decided to put off his initial choice of study and join Wundt's psychological laboratory instead. This decision marked the starting point of a very successful career. Thanks to the support of William James, Harvard would be his academic home base from the early 1890s onward. By the end of this decade, he was elected president of the American Psychological Association.[1] Wundt had good reason to proudly celebrate the successes of the young man who had joined his institute fifteen years before.

The correspondence between the successful student and his former teacher, however, bears witness to a tense relationship. As early as 1890, Münsterberg's letters show that he was burdened by the suspicion that Wundt did not appreciate him all that much. Friends and acquaintances had told him that his *Doktorvater* had privately asserted that his former student had been ungrateful to him. In a letter to Wundt, Münsterberg sadly claimed that, while he could accept his former teacher's criticism of his scholarly accomplishments, he would "lose [his] self-respect when

© The Author(s), under exclusive license to Springer Nature Switzerland AG 2022
C. Engberts, *Scholarly Virtues in Nineteenth-Century Sciences and Humanities*, https://doi.org/10.1007/978-3-030-84566-7_1

[these] accusations of ungratefulness would be warranted."[2] He pointed out that he continued to express his indebtedness to Wundt in many ways: "I ostentatiously present myself as your student toward all your detractors; a framed picture of you is the only decoration on the walls of my laboratory; I sent you the first copy of all my books." These displays of loyalty did not, however, appease Wundt.

Wundt indignantly denied that he had ever accused Münsterberg of ingratitude and took on the offensive. He wryly asserted that Münsterberg's recent work had been "rushed and not sufficiently matured."[3] He added insult to injury by pointing out that his students could only display their gratitude by working "reliably, diligently, and meticulously without caring about authorities or [their] career." He essentially argued that the rushed and immature character of Münsterberg's latest publications should be interpreted as the regrettable neglect to display gratitude in an appropriate and convincing manner. Wundt's addition that he had shared his unfavourable opinion of his former student's recent work with mutual acquaintances suggests that the worrying reports of Münsterberg's friends may have been accurate.

Though Wundt and Münsterberg would stay in touch until the latter died in 1916, the pattern of mutual recrimination established in the early 1890s would continue to shape their relationship. Münsterberg referred to the "rushed production" of his early career over and over again in his letters.[4] He also continued to try to convince his *Doktorvater* of his deeply felt gratitude. One letter contained an extensive apology for all the ways in which he had "knowingly or unknowingly hurt or wronged" Wundt and an additional excuse for all the things for which he had "not sufficiently expressed his sincere gratitude."[5] A few month later, he even asked him to recognize him as his "most grateful student."[6] At the opening of his new psychological laboratory at Harvard in 1905, he stressed his gratitude again: "I just wanted to say [...] that today I am profoundly aware of my dependence on and my gratitude for you and that I am [...] guided by the desire that this workplace will be imbued with your spirit."[7] In 1915, he even wrote his former teacher that he had started a campaign to get him a Nobel Prize.[8]

For the purposes of this study, the significance of these awkward exchanges between Wundt and Münsterberg lies not in the merit of their work. I will neither look into the supposed hasty sloppiness of the student's early work, nor will I discuss whether the work of the teacher would have merited a Nobel Prize. Their correspondence is mostly of interest to me because it provides a clear example of scholars employing a language of virtue to discuss both their work and their relationship. Both men take it for granted that a student's gratitude is essential to maintaining a good relationship with his former teacher, and both agree that good scholarship is the result of virtues like reliability, diligence, and meticulousness. Their correspondence also suggests that such virtues cannot be understood in isolation from each other.[9] In Wundt's eyes, gratefulness is clearly bound up with meticulousness and other virtues, and Münsterberg does not challenge this assumption. This study investigates this interwovenness of broadly shared virtues of scholarship on the basis of in-depth case studies from nineteenth and early twentieth-century Germany.

Loyalty and Independence

The virtues at the heart of this study are critical independence and collegial loyalty. Loyalty and independence shed a particularly bright light on the complex entanglement of separate virtues because—as a wide range of case studies will illustrate—they touch on each other in a lot of different ways. Sometimes a sense of mutual loyalty can discourage scholars from criticizing each other's work or character. At other times, a similar sense of loyalty might contribute to an atmosphere in which scholars are encouraged to share well-thought-out and highly useful critiques of their peers' efforts. Ideas about loyalty and independence are also shaped by the specific circumstances in which these virtues become relevant: loyalty towards a colleague can not necessarily be understood in the same way as loyalty towards a whole scholarly community, and the desire to be independent from a faculty board might not carry the same weight as ideals about teaching and the proper relation between a professor and his students. Finally, sometimes the strongest tensions might not be between loyalty and independence as such but between loyalties towards a myriad of people and institutions.

Another reason to concentrate on loyalty and independence is the fact that this focus allows for an understanding of the morality of scholarly life as shaped by some of the same values and social structures that shape whole societies. Both loyalty and independence have been identified as key elements of nineteenth-century bourgeois morality. Lothar Gall has argued that shared conceptions of "spiritual and moral autonomy" as well as "practical independence" were central to the identity of nineteenth-century, middle-class Germans.[10] Manfred Hettling also draws attention to the bourgeois character of the widespread expectation that middle-class men should determine their own life goals.[11] This appreciation of middle-class autonomy is emphasized in Deirdre McCloskey's outline of the Western bourgeois virtues as well.[12] She also adds loyalty to this list of virtues. The middle-class connotations of loyalty are also highlighted by Ute Frevert and Ulrich Schreiterer, who argue that it acquired a "specific bourgeois colouring" in nineteenth-century Germany.[13]

Loyalty and independence were not only shaped by class structures: conceptions of gender also moulded the way in which these virtues were understood. Middle-class men conceived of loyalty as an element of male friendship and camaraderie. Robert Nye understands this bourgeois culture as a continuation and appropriation of a late medieval honour culture in which "unambiguous demonstrations of loyalty were regarded as the highest expression of a man's virtue."[14] Nicolaus Sombart emphasizes that in nineteenth-century Germany, friendship was "a male preserve just like the 'state' is a male preserve (only men can be friends)."[15] The expectation that the nineteenth-century university would be a space for male sociability was also unambiguously articulated by the Berlin legal scholar who stated that "our universities are universities for men" and strictly "tailored to the male spirit (*Geist*)."[16] This was not unique to German academia: Paul Deslandes argues that British universities were "highly gendered little worlds characterized by intense institutional loyalty [...] and carefully articulated visions of male solidarity."[17]

Nineteenth-century conceptions of independence also emphasized the gendered character of academia. This was entirely in line with the broader intellectual trend to acknowledge male independence without recognizing the same quality in women.[18] German scholars often doubted women's ability to independently contribute to scholarship. A Strasbourg gynaecologist is quite typical in his condescending assessment: "Usually [women's] scientific accomplishments do not rise above the level of mediocrity. Never has a woman set herself a grand, scientific task; never has she succeeded in solving even an easy task in a ground-breaking way."[19] Like the gendered conception of loyalty, the close connection between independence and masculinity was not unique to the German lands. Matthew McCormack has argued that the "Georgian man's sense of gender" was often articulated in terms of his "manly independence."[20] A similar conception of masculine autonomy is central to Hannah and John Gay's description of the way in which ideals of manliness shaped the ideals of good scholarship in nineteenth-century Britain.[21] Furthermore, Robert

Nye points out that assumptions of masculine independence were widely shared among contemporary French physicians and scientists as well.[22]

Scholarly Virtues and the History of Scholarship

Until recently, historians of science have only paid limited attention to scholarly virtues. Steven Shapin notes that twentieth-century scholars usually present the history of scholarship as an impersonal process that is best understood by highlighting the development of "rationally organised and regulated institutions."[23] He bemoans the fact that this approach hides the fact that "at least since the seventeenth century, familiar people and their virtues have always been pertinent to the making, maintenance, transmission and authority of knowledge."[24] More recently, however, a growing number of authors has tried to meet this challenge by investigating questions of virtue and vice in the history of scholarship. From the late 1980s onward, these scholars have produced a highly diverse body of work. Their studies of scholarly virtue have provided fertile ground for new questions about the history of scholarly in the twenty-first century.[25]

Shapin himself was one of the first historians of science to demonstrate the promise of investigating scholarly virtue. In his reflections on credibility in early modern science, he found that virtue and trustworthiness were intimately connected. Only a scientist with a virtuous reputation was able to convince others of the truth and the significance of his findings. A reputation for accuracy and love of truth—two virtues generally associated with men rather than women—were indispensable in persuading one's peers.[26] Trust was not only dependent on perceived virtue among Shapin's early modern British scholars. More recently, Kasper Eskildsen has made similar observations about nineteenth-century German historians, who carefully conveyed their trustworthiness by making sure that their work would draw attention to their accuracy, honesty, and impartiality.[27]

The growing interest in scholarly virtue and vice has also inspired early modernists who study the complex relation between religion and science. Peter Harrison has investigated the changing conception of curiosity. In the writings of the church fathers, curiosity was presented as a vice but from the seventeenth century onward it was increasingly recognized as an epistemic virtue.[28] Michael Heyd's study of the changing early modern responses to religiously inspired enthusiasm further contributes to a conception of the seventeenth century as a time during which the relation between religion, science, and scientific virtue was thoroughly and decisively reconsidered.[29] Sari Kivistö's study of the vices of learning, as recognized at early modern universities, also draws attention to the way in which appeals to "the importance of traditional moral and religious values" have shaped "conflicting notions of knowledge and scholarly ethics."[30]

Other recent engagements with scholarly virtues and vices are primarily grounded in an interest in the philosophy of science. At least two common philosophical approaches are inspired by this line of thought. One group of philosophers is mainly interested in what is often called "virtue epistemology" and builds on the assumption that "in one way or another all virtues have a cognitive aspect."[31] Other virtue epistemologists conceptualize intellectual virtues as forms of moral virtue.[32] A second group of philosophers interested in questions of virtue and vice primarily concentrates on the discussion of scientific research ethics. Some of these scholars opt for a descriptive and historicizing approach. Albert Jonsen's and Robert Baker's investigations into the historical roots of contemporary medical ethics in the United States provide an example of this genre.[33] Inspired by Alasdair MacIntyre's seminal *After Virtue*, others engage in a more prescriptive endeavour and aim to develop more practical guidelines on the basis of a virtue ethical approach to scientific research ethics.[34]

This book is not a follow-up to this growing body of literature inspired by interests in religion and philosophy. It does, however, build on another promising feature of the study of scholarly virtues that has recently been highlighted by several authors: its suitability as a starting point for cross-disciplinary comparison. In their introduction to a recent volume about epistemic virtues, Jeroen van Dongen and Herman Paul, for example, argue that this is a promising field of investigation exactly because it has the potential to contribute to a history of scholarship that goes beyond the usual focus on the development of specific disciplines.[35] This approach reflects an increasing interest in the cross-disciplinary study of the history of scholarship and allows for meaningful comparisons between the sciences and the humanities.[36] In his famous essay on the "two cultures," C.P. Snow argued that the sciences and the humanities "are so different that, even on the level of emotion they can't find much common ground."[37] A careful examination of the widely recognized virtues of critical independence and collegial loyalty, however, offers a way in which meaningful comparisons are very well possible.

Moral Economies of Scholarship

The relation between loyalty and independence is shaped by a myriad of variables. From the perspective of the individual scholar, the relative weight of both virtues might change over time. The relative importance of both virtues might also be shaped by the character of his or her relationship to individual peers and institutions: not everyone inspires the same degree of loyalty or urge to assert independence. In this study, I will use the metaphor of a moral economy of scholarship to shed light on the complex and ever-changing relationship between virtues. I use the word "economy" not to invoke the image of some sort of moral marketplace in which scholars compete for their reputation. Instead, I draw on recent discussions of the history of science in which the notion of moral economies is used to underline the extent to which the social nature of everyday scholarly work—in both the sciences and the humanities—contributes to a dynamic relation between widely shared virtues.

In his study of early twentieth-century fruit fly geneticists, Robert Kohler was one of the first historians of science to refer to a moral economy of scholarship.[38] He gave a detailed description of the virtues that shaped the collaboration of a group of innovative researchers. This group was defined not only by shared research interests but also by their employment at a limited number of interconnected laboratories during a relatively short period of time. This clearly defined case selection allowed Kohler to give a detailed and contextualized description of the virtue assessments of the researchers to whom he referred to as the "fly people." He concluded that their moral economy amounted to "a moral ethos of cooperation and communality" that resulted from the abundance of research possibilities suggested by the new genetics research program and the close personal relations between its most ardent advocates.[39] Kohler's emphasis on cooperation and communality illustrates that the moral economy metaphor does not have to be understood as a reference to scholars competing in some sort of moral marketplace of scholarship. At the same time, however, his understanding of the term "moral economy" is not entirely clear because he does not explicitly reflect on its usefulness or definition.

Shortly after the publication of Kohler's study, Lorraine Daston discussed the idea of a moral economy of scholarship from a more conceptual point of view. She follows Kohler in emphasizing that the word "economy" does not refer to the "money, markets, labor, production, and distribution of material resources, but rather to an organized system that displays certain regularities."[40] She then defines a moral economy as a "web of affect-saturated values that stand and function in well-defined relationship to one another" with "equilibrium points and constraints."[41] In this study, I will build on Daston's definition to the extent that I am primarily interested in the relationship between different values, especially critical independence and collegial loyalty, and the wide variety of equilibriums that emerges in the everyday evaluative practices encouraged by the quintessentially social character of scholarship.

The focus on relations and equilibriums allows me to make a new contribution to current debates about epistemic and scholarly virtues. Recent studies are usually limited to the discussion of single virtues. Peter Harrison's study into the virtue of curiosity and Herman Paul's reflections on loyalty are though-provoking examples of this growing body of literature.[42] Lorraine Daston's and Peter Galison's ground-breaking study of objectivity and Kathryn Murphy's and Anita Traninger's edited volume about impartiality also share this emphasis on a single virtue.[43] Sari Kivistö's study of the moral environment of early modern universities discusses multiple vices, such as self-love and the desire for fame, but does not look systemically at the dynamics of the relations between single virtues and vices.[44] One of this study's main contributions to the existing literature can therefore be found in its emphasis on the way in which scholars attempt to find a balance between the—sometimes conflicting and sometimes overlapping—demands of coexisting virtues.

Significantly, the virtues central to this investigation, critical independence and collegial loyalty, not only are relevant to individual scholars' research endeavours but also decisively shape their social relations. The social function of these specific virtues ensures that they are reflected in—and reproduced by—a wide variety of means of scholarly communication, such as private letters, published works, and bureaucratic correspondence. This omnipresence makes the analysis of independence and loyalty uniquely suited to shed light on the tensions and associations between single virtues. In its emphasis on these specific virtues, this study also adds to the discussions of scholarly virtues by Daston, Galison, Murphy, Traninger, and others. While their focus on virtues like objectivity and impartiality invites an analysis of the relation between researchers and the natural world, a closer look at the virtues of independence and loyalty contributes to a better understanding of the quintessentially social nature of scholarship.

Science and Scholarship in Germany and Abroad

The study of the virtues of independence and loyalty does not only invite an investigation across disciplinary borders and a close look at the way in which the metaphor of the moral economy of scholarship can shed light on the ever-changing balance between single virtues. A quick glance at a variety of accounts of the history of scholarship suggests that the recognition of the importance of these virtues also adds to our notions of the development of science in different countries and in different times. A few short examples will illustrate the way in which these virtues can help us to gain a deeper understanding of some of the most widely discussed episodes in the history of Western scholarship.

The personal entanglements that led to the completion of Isaac Newton's *Principia* (*The Mathematical Principles of Natural Philosophy*) provide a first striking example. The publication of this influential study was made possible by the collaboration between Newton and Edmund Halley—a collaboration that was characterized by Halley's loyal support of Newton as well as by his critical involvement with the ambitious project.[45] It was set in motion by Halley's consultation of Newton about the trajectories of celestial bodies. He then urged him to further develop his nonchalant but far from self-evident reply that it would have to be an ellipse. In the subsequent years, Halley provided astronomical data, convinced the Royal Society to support Newton's study of these data, and paid for the eventual publication of the *Principia*. Last but not least, he also recalculated and corrected all the measurements on the basis of which Newton had calculated celestial orbits. On the basis of these critical assessments, the measurements of the trajectory of the comet observed in 1680—today known as Halley's comet—were corrected. Without Halley's loyal support and critical involvement, the *Principia* would have been a quite different study, if it would have been realized at all!

Loyalty and independence also play a central role in common understandings of another chapter in the history of science that took place two centuries later at the other side of the English Channel: the combined accomplishments of Marie and Pierre Curie. Their close collaboration on the study of the radioactivity of two minerals of uranium led to the discovery of two new elements: radium and polonium. In an environment in which women were not expected to be able to make such novel contributions to scientific knowledge, it was not easy to convince the academic world of Marie's independent input: some claimed that Pierre had done all the important work, while Marie had simply clung to his coattails.[46] Her husband recognized the importance that their scientific peers attached to supposedly independent contributions to scholarship and loyally acknowledged the indispensability of his wife's contributions to their collaborative efforts.[47] He did not, however, convince all of his colleagues: almost two decades after Pierre's death, Marie still had to go "to great lengths to distinguish who did what" in her biography of her husband.[48]

At the same time, the widely shared acknowledgement of scholarly loyalty was also clearly visible in the book genre that has gone by such different names as *Festschrift*, homage literature, or *liber amicorum*. These books are volumes of papers in the honour of an individual scholar, most often compiled to celebrate their sixtieth or seventieth birthday, or retirement. The roots of this genre can be found in nineteenth-century Germany, but it soon caught on in other European countries and the United States, as well.[49] In the late nineteenth and early twentieth centuries, contributors to *Festschriften* usually refrained from criticism of the scholar to be honoured because the main purpose of the volume was to draw attention to the successes and continued influence and relevance of the *Festschrift*'s recipient.[50] Even though the scholarly merit of these volumes is occasionally disputed—some have called it "a noxious weed to be uprooted"—only very few academics would decline an invitation to contribute to one even today.[51] Refusing to contribute can all too easily be interpreted as regrettable disloyalty towards a valued colleague or former teacher.

The history of the homage volumes also points towards one of the main reasons why this study is based on German case studies. Both early nineteenth-century Prussian government administrators and their successors in the German *Reich* successfully promoted academic research and university education. Their efforts are exemplified by the early nineteenth-century innovations of Wilhelm von Humboldt and *Ministerialdirektor* Friedrich Althoff's close involvement with almost all aspects of university governance at the end of the century.[52] One indicator of the success of the German state's supportive involvement with university affairs is the fact that by the early twentieth century, German had become the world's most important language of scientific publishing. The decline of the prominence of German as a global academic language only started after the First World War and was accelerated by the rapid development of English into the new *lingua franca* of international scholarship.[53]

All of the most significant developments in the organization of Western scholarship occurred in German academia as well. These developments included—but were not limited to—the establishment of scholarly societies, the founding of an impressive number of new scientific journals, the increasingly frequent collaboration between industry and academia, and the continuous growth of the number of students and university teachers.[54] Some other novel developments of nineteenth-century academic life across the Western world even originated in Germany. The emphasis on research in the educational mission of the newly established Johns Hopkins University, for instance, was inspired by German examples of graduate education.[55] At the same time, the interest in German models of university management also encouraged other American institutions of higher education to develop into research universities. At all these institutions, the training of graduate students and the conduct of research by staff members became increasingly important.[56]

This pervasive German influence abroad was observable both in the novel ways in which university education was organized and in the high number of staff members who were decisively shaped by their personal interactions with German scholars. One educational innovation that was widely copied was the seminar. The roots of seminar education can be traced to eighteenth-century Germany.[57] The Humboldtian university reforms and a renewed interest in scientific education in the German Empire in the 1870s turned seminar meetings into a structural feature of university teaching that was soon copied all over the world.[58] The research-oriented approach to education that was cultivated in the seminars of the humanities faculties was also adapted to education in the sciences: laboratories became increasingly important as centres of education.[59] The men who established these new laboratories and seminars all over the world often had first-hand experience with seminar or laboratory work in Germany. An incomplete list of the psychological laboratories founded in America by students of Wundt illustrates this observation: James McKeen Cattell founded the laboratories of the University of Pennsylvania and Columbia University, Frank Angell those of the Cornell and Stanford Universities, Edward W. Scripture established the psychological laboratory of Yale, and George Stratton did the same at the University of California, Berkeley.[60]

Three Disciplines

In this study, I engage with conceptual questions about the relation between scholarly virtues through the lens of everyday practices of scholarly life. I will primarily focus on evaluative practices because these are particularly suited to shed light questions of loyalty and independence. After all, evaluations offer a unique opportunity to either assert independence from one's peers through criticism or to show one's collegial loyalty through public praise or constructive private feedback. The evaluative practices that I will most extensively discuss are private correspondences, the editing of journals aimed at a scholarly audience, book reviews, and letters of recommendation.

The most insightful way to get a clear picture of the impact of different evaluative practices on individual scholars' efforts to balance virtues is to focus on the professional networks of a small number of scholars. This allows me to reflect on, for example, a scholar's private relationships in the light of his highly critical book reviews or to understand his editorial decisions in the light of his personal relation with the contributors to this journal. Therefore, this study will focus on evaluative practices in the professional network of one humanities scholar, one scientist, and one academic who would today be categorized as a social scientist. The protagonists in question are the orientalist Theodor Nöldeke, a small network of researchers associated with the research institutes of the bacteriologist Robert Koch, and the aforementioned experimental psychologist Wilhelm Wundt.

I have selected scholars from these three disciplines because this allows for a cross-disciplinary analysis of loyalty and independence. They have also been chosen because scholars in each of these disciplines were confronted with questions about the future shape of their fields of research. These processes of redefinition allowed ample room for discussion and disagreement and encouraged scholars to reflect on their relations both to each other and to the institutions that facilitated their research. In psychology and bacteriology, these debates resulted from the fact that both fields of research were relatively new. One fundamental question that occupied psychologists' minds was which mental processes could and which could not be studied through experimental methods.[61] Another debate concerned the institutional home of this new field of research. Some early psychologists preferred to remain part of the philosophy departments in which the first psychological chairs were created. They had to defend themselves, however, against criticisms of both psychologists and other members of the philosophy departments. An increasing number of psychologists wanted a new institutional home for their new discipline, while other philosophers often resented the fact that they had to compete with psychologists for scarce resources and appointments.[62]

The newness of bacteriology did not lead to similar questions of institutional belonging. The main challenge for bacteriologists was to overcome the scepticism of both medical practitioners and the general public. The doubts of the sceptics were confirmed in at least one widely covered case in which Robert Koch himself had played a central role.[63] Popular distrust of bacteriological findings was further reinforced by the large sums of money that could be made through the sale of newly developed pharmaceuticals. Bacteriologists were therefore vulnerable to the accusation that they were merely self-serving entrepreneurs.[64] The relatively new technique of vaccination—one of the major innovations of the new bacteriological discipline—also had to deal with the stubborn resistance of a small but vocal group of members of the medical establishment, general practitioners, and concerned citizens.[65]

The debates among orientalists did not result from the newness of the discipline: the study of oriental languages at European universities has a long history. By the end of the nineteenth century, however, new fault lines were becoming visible within the discipline. It became increasingly difficult, for example, to find common ground between an older generation of theologically inspired Arabists and a younger generation that advocated a more secular approach to the study of the Middle East.[66] Another debate between the generations concerned research priorities. Traditional orientalists usually limited themselves to the study of ancient languages and texts, while their younger peers increasingly directed their attention towards modern languages and contemporary culture. This shifting research focus was particularly controversial because it developed hand in hand with Europe's colonial endeavours and therefore posed a challenge to the old—philologically inclined—guard's self-image of scholarly disinterestedness.[67]

The Protagonists and Their Networks

Theodor Nöldeke, on whom I my investigation into the evaluative practices among orientalists is focused, had a wide network of scholars from different generations and convictions. His *Doktorvater* was Heinrich

Ewald, whose studies were an inextricable mixture of theology and linguistics, but he had a more amiable relation with Ewald's less theologically inclined contemporary, the Leipzig Arabist Heinrich Leberecht Fleischer.[68] The philological inclinations of Fleischer were markedly different from the interest in contemporary languages and culture among Nöldeke's closest colleagues of a younger generation, such as the Dutch Arabist Christiaan Snouck Hurgronje and his German colleague Carl Heinrich Becker.[69] His network was further expanded and maintained by his membership and board membership of the *Deutsche Morgenländische Gesellschaft* (*DMG*), the national association of German Orientalists. His network was not limited to Germany. He was, for example, also an important member of the international consortium assembled by his Leiden friend and colleague Michael Jan de Goeje for his long-running project to edit the full *Annals* of al-Tabari.[70] In addition, he was a very productive writer of book reviews: in the 1870s, for example, he wrote more than a hundred reviews for the *Literarisches Centralblatt für Deutschland*.[71]

During this same period, Wilhelm Wundt, the psychologist whose network and evaluative practices are as central to this study as Nöldeke's, also wrote close to a hundred reviews for the *Centralblatt*. As a true pioneer in his chosen field of experimental psychology, Wundt did not maintain particularly close relations with scholars of the earlier generation, even though he developed an amiable relationship with a kindred spirit of this older peer group, the psychophysicist Gustav Theodor Fechner.[72] The heart of Wundt's personal network can be traced back to the Institute for Experimental Psychology that he established in Leipzig in 1879. An impressive number of influential scholars were trained in here, including the psychiatrist Emil Kraepelin, the psychologist Oswald Külpe, the pedagogical scholar Ernest Meumann, and the large number of influential American psychologist mentioned earlier. This laboratory was also the institutional home of the *Philosophische Studien*, a scholarly journal edited by Wundt himself.[73]

The other network at the heart of this study can be traced back to Berlin's *Institut für Infektionskrankheiten* (Institute for Infectious Diseases) in the early 1890s. The institute was founded to support the research of Robert Koch in the wake of his claims to have invented a cure for tuberculosis. He surrounded himself with a large number of talented young researchers who worked on prophylactics and cures for a wide variety of infectious diseases. Two of his most successful early collaborators were Emil Behring and Paul Ehrlich. Both men would eventually leave the institute to pursue independent careers. From this moment on, their relations—especially between Behring and his former colleagues—grew increasingly tense, even though they would remain in touch until the end of their careers. Their continued contact was not to a small degree facilitated by *Ministerialdirektor* Friedrich Althoff at the Prussian Ministry of Education. Among other things, the interactions of Koch, Behring, and Ehrlich with this government official provide an insightful glimpse at the intricacies of university appointments.[74]

This primary cast of protagonists will be supplemented with a number of scholars whose contributions are less central to this study's aim to provide a multidisciplinary comparison. Two of them play such an important role in my argument that they deserve an early mention as well. First, the policies and networks of the aforementioned Friedrich Althoff will be discussed at some length. These are particularly relevant to the comparative character of this study because in his capacity as *Ministerialdirektor*, Althoff was involved with a wide range of decisions that shaped the institutional character and individual careers in both the sciences and the humanities.[75] A second man who will feature quite prominently in this study is Friedrich Zarncke, a Leipzig Germanist and the editor of the *Literarisches Centralblatt*. Because scholars from different disciplines—among them both Nöldeke and Wundt—generously contributed to his widely read review journal, his correspondence sheds light on the activities and expectations of an editor as well as the norms and values promoted and maintained in late nineteenth-century book reviews.[76]

A Quick Look Ahead

As mentioned before, the notion of evaluative practices is central to my presentation of a moral economy characterized by loyal collegiality and critical independence. Each of the following chapters will therefore be dedicated to one specific evaluative practice and present an investigation of the ways in which this particular practice shaped the ever-changing balance between loyalty and independence in the context of the everyday working life of scholars in different disciplines.

Chapter 2 explores practices of informal mutual evaluation. In all disciplines, scholars commented on each other's work before it would be shared with a wider audience. This informal, prepublication feedback could take different shapes, such as the proofreading of books, articles, and text editions as well as the extensive testing of newly developed medical substances. Among other things, this chapter demonstrates the way in which personal trust and loyalty created the necessary conditions that allowed for honest and straightforward independent criticism. The chapter also shows, however, that there are remarkable differences between disciplines. Informal evaluation among orientalists is shaped by the solitary character of most of their endeavours, while evaluative practices in experimental psychology are moulded by the social environment of the laboratory. Bacteriologists' attitudes towards mutual evaluation are strongly influenced by the general public's hopes and expectations about the efficacy and safety of the fruits of their labour.

Chapter 3 provides a close look at the practices of editing a scholarly journal. I will explore these practices through a close look at the editorial work at three journals: Wundt's *Philosophische Studien*, the *Zeitschrift der Deutschen Morgenländischen Gesellschaft*, and Zarncke's *Literarisches Centralblatt*. In none of these cases the editor's task was simply selecting the contributions with the highest scholarly merit. Editorial decisions were also shaped by the interests of a wide range of stakeholders that includes publishers, contributors, and the intended audience. I will further argue that these journals' different editorial policies are best understood not on the basis of disciplinary differences but on the basis of other characteristics of each publication. The *Studien* were edited by one all-powerful editor who did not have to worry about profitability, the editorial board of the *Zeitschrift* was bounded by the fact that their journal was the official organ of a society that aimed to represent all German orientalists, and the *Centralblatt* was engaged in a constant struggle to make ends meet financially.

In Chap. 4, I follow the lead of David Shatz who has stated that "readers of published book reviews are given various kinds of judgements that they do not find in scholarly articles that critique other people's work. How good is the writing style? Is the prose lucid? Is the work well organised? Are there typos?"[77] I will take a closer look at Zarncke's *Centralblatt* through a detailed analysis of the book reviews that Nöldeke and Wundt contributed to his journal in the 1870s. I will argue that book reviews were not only used to evaluate each other's work but also to evaluate each other's character and to create, maintain, and police a scholarly community characterized by a supposedly shared commitment to certain ideals of good scholarship. A comparison between the virtues and skills expected across disciplines shows not only a lot of similarities but also some striking differences.

Chapter 5 is an exploration of the letters of recommendation for university appointments, most of which can be found in the papers of Friedrich Althoff. A close look at these letters shows that hiring decisions were only to a limited extent based on the sort of ideals of academic excellence that can be demonstrated through long lists of widely read scholarly publications. Qualities that appear to be as consequential as publication records include class, sociability, teaching prowess, and managerial abilities. This last quality also suggests that there were indeed differences between the requirements in different disciplines. Managerial skills were particularly important in disciplines characterized by intense collaboration in, for example, a clinic or a laboratory. Candidates for a position at a philosophical institute, on the other hand, were more thoroughly screened for their personal character and the moral and religious qualities expressed in their teaching and writing.

In the final full chapter, I will bring together the findings of the previous case studies. I reflect on the ways in which the performance of critical independence and collegial loyalty in different evaluative practices amounts to a moral economy of scholarship. I pay attention to similarities and differences across disciplines as well as to the way in which this moral economy shapes a wide variety of relationships between individual scholars, academic institutions, and the outside world. Because my findings about nineteenth-century Germany invite comparison to modern-day conceptions and performances of virtuous scholarship, I reflect on the relevance of these past ideals to contemporary academic life in the epilogue. After all, most of the practices that I describe still exist today and the virtues of loyalty and independence are still valued by twenty-first-century scholars. A better understanding of the moral economy of the past might help us to make better sense of the expectations and values that shape today's academic life as well.

Notes

1. An overview of Münsterberg's life and career is given in: Phyllis Keller, *States of belonging: German-American intellectuals and the First World War* (Cambridge, MA: Harvard University Press, 1979), 5–118.
2. Hugo Münsterberg to Wilhelm Wundt, 10 November 1890, Universitätsarchiv Leipzig (hereafter UAL), NA Wundt/III/701–8000/764b/415–426.

3. Wilhelm Wundt to Hugo Münsterberg, 12 November 1890, UAL, NA Wundt/III/701–800/765/427–438.
4. Hugo Münsterberg to Wilhelm Wundt, 31 March 1896, UAL, NA Wundt/III/701–800/765f/501–512 and 5 November 1905, UAL, NA Wundt/III/701–800/768a/607–622.
5. Hugo Münsterberg to Wilhelm Wundt, 7 April 1896, UAL, NA Wundt/III/701–800/765g/513–528.
6. Hugo Münsterberg to Wilhelm Wundt, 14 April 1896, UAL, NA Wundt/III/701–800/765h/529–532.
7. Hugo Münsterberg to Wilhelm Wundt, 5 November 1905, UAL, NA Wundt/III/701–800/768a/607–622.
8. Hugo Münsterberg to Wilhelm Wundt, 20 October 1915, UAL: NA Wundt/III/701–800/768e/663–670.
9. I discuss the relation between Wundt and Münsterberg at more length in Christiaan Engberts, "Wilhelm Wundt's Critical Loyalty: Balancing Gendered Virtues Among Early Experimental Psychologists," in *Gender, Embodiment, and the History of the Scholarly Persona: Incarnations and Contestation*, ed. Kirsti Niskanen and Michael J. Barany (London: Palgrave Macmillan, 2021), 287–314.
10. Lothar Gall, *Bürgertum, liberale Bewegung und Nation: Ausgewählte Aufsätze*, edited by Dieter Hein, Andreas Schulz and Eckhardt Treichel (Munich: Oldenbourg, 1996), 12.
11. Manfred Hettling, "Die persönliche Selbständigkeit: Der archimedischer Punkt bürgerlicher Lebensführung," in *Der bürgerliche Wertehimmel: Innenansichten des 19. Jahrhunderts*, ed. Manfred Hettling and Stefan-Ludwig Hoffmann (Göttingen: Vandenhoeck and Ruprecht, 2000), 57.
12. Deirdre N. McCloskey, *The Bourgeois Virtues: Ethics for an Age of Commerce* (Chicago, IL: University of Chicago Press, 2006), 66.
13. Ute Frevert and Ulricht Schreiterer, "Treu: Ansichten des 19. Jahrhunderts," in *Der bürgerliche Wertehimmel: Innenansichten des 19. Jahrhunderts*, ed. Manfred Hettling and Stefan-Ludwid Hoffmann (Göttingen: Vandenhoeck and Ruprecht, 2000), 219–220.
14. Robert A. Nye, "Medicine and Science as Masculine 'Fields of Honor'," *Osiris* 12 (1997): 61.
15. Nicolaus Sombart, "Männerbund und Politische Kultur in Deutschland," in *Männergeschichte—Geschlechtergeschichte: Männlichkeit im Wandel der Moderne*, ed. Thomas Kühne (Frankfurt: Campus, 1996), 137.
16. Otto Gierke in *Die Akademische Frau. Gutachten hervorragender Universitätsprofessoren, Frauenlehrer und Schriftsteller über die Befähigung der Frau zum wissenschaftlichen Studium und Berufe*, ed. Arthur Kirchoff (Berlin: Hugo Steinitz, 1897), 23.

17. Paul R. Deslandes, *British Masculinity and the Undergraduate Experience, 1850–1920* (Bloomington, IN: Indiana University Press, 2005), 18.
18. Hettling, "Die persönliche Selbständigkeit," 70–71.
19. Wilhem Alex Freund in: Kirchhoff, *Die Akademische Frau*, 181.
20. Matthew McCormack, *The Independent Man: Citizenship and Gender Politics in Georgian England* (Manchester: Manchester University Press, 2005), 2.
21. Hannah Gay and John W. Gay, "Brother in Science: Science and Fraternal Culture in Nineteenth-Century Britain," *History of Science* 35, no. 4 (1997): 427.
22. Nye, "Medicine and Science," 75.
23. Steven Shapin, *The Scientific Life: A Moral History of a Late Modern Vocation* (Chicago, IL: Chicago University Press, 2008), 3.
24. Ibid., 4.
25. In the epilogue of this study, I reflect on questions the relevance of scholarly virtues to the understanding of modern-day scholarship.
26. Steven Shapin, "The House of Experiment in Seventeenth-Century England," *Isis* 79, no. 3 (1988): 397–398; Steven Shapin, *A Social History of Truth: Civility and Science in Seventeenth-Century England* (Chicago, IL: University of Chicago Press, 1994), 89.
27. Kasper Risbjerg Eskildsen, "Inventing the archive: Testimony and virtue in modern historiography," *History of the Human Sciences* 26, no. 4 (2013); 11.
28. Peter Harrison, "Curiosity, Forbidden Knowledge, and the Reformation of Natural Philosophy in Early Modern England," *Isis* 92, no. 2 (2001); 267, 283.
29. Michael Heyd, *"Be Sober and Reasonable": The Critique of Enthusiasm in the Seventeenth and Early Eighteenth Centuries* (Leiden: Brill, 1995).
30. Sari Kivistö, *The Vices of Learning: Morality and Knowledge at Early Modern Universities* (Leiden: Brill, 2014), 13.
31. Robert C. Roberts and W. Jay Wood, *Intellectual Virtues: An Essay into Regulative Epistemology* (Oxford: Clarendon Press, 2007), 59.
32. Linda Trinkaus Zagzebski, *Virtues of the mind: An inquiry into the nature of virtue and the ethical foundations of knowledge* (Cambridge: Cambridge University Press, 1996), xiv.
33. Robert Baker, *Before Bioethics: A History of American Medical Ethics from the Colonial Period to the Bioethics Revolution* (Oxford: Oxford University Press, 2013); Albert R. Jonsen, *The Birth of Bioethics* (Oxford: Oxford University Press, 1998).
34. Alasdair MacIntyre, *After Virtue: A Study in Moral Theory* (London: Duckworth, 1981); Robert T. Pennock and Michael O'Rourke,

"Developing a Scientific Virtue-Based Approach to Science Ethics Training," *Science and Engineering Ethics* 23, no. 1 (2017).

35. Jeroen van Dongen and Herman Paul, "Introduction: Epistemic Virtues in the Sciences and the Humanities," in *Epistemic Virtues in the Sciences and the Humanities*, ed. Jeroen van Dongen and Herman Paul (Cham: Springer, 2017), 5.
36. See, for example: Rens Bod, *A New History of the Humanities: The Search for Principles and Patterns from Antiquity to the Present* (Oxford: Oxford University Press, 2013).
37. C.P. Snow, *The Two Cultures and the Scientific Revolution* (New York: Cambridge University Press, 1959).
38. Robert E. Kohler, *Lords of the Fly: Drosophila Genetics and the Experimental Life* (Chicago, IL: University of Chicago Press, 1994).
39. Ibid., 92–93.
40. Lorraine Daston, "The Moral Economy of Science," *Osiris* 10 (1995): 4.
41. Ibid.
42. Harrison, "Curiosity, Forbidden Knowledge, and the Reformation" and Herman Paul, "Germanic Loyalty in Nineteenth-Century Historical Studies: A Multi-Layered Virtue," *História da Historiografia* 12, no. 30 (2019).
43. Lorraine Daston and Peter Galison, *Objectivity* (New York, NY: Zone Books, 2007) and Kathryn Murphy and Anita Traninger (eds.), *The Emergence of Impartiality* (Leiden: Brill, 2014).
44. Kivistö, *The Vices of Learning*.
45. In this section I follow: Lisa Jardine, *Ingenious Pursuits: Building the Scientific Revolution* (London: Little, Brown and Company, 1999), 11–41.
46. Naomi Pasachoff, *Marie Curie: And the Science of Radioactivity* (Oxford: Oxford University Press, 1996), 70.
47. Ibid., 52.
48. Eva Hemmungs Wirtén, *Making Marie Curie: Intellectual Property and Celebrity Culture in an Age of Information* (Chicago, IL: Chicago University Press, 2015), 28.
49. Alice P. Kenney and Leslie J. Workman, "Volumes of Homage: Festschriften in America," *New York History* 55, no. 4 (1974): 460–462.
50. Andrea Ender and Bernhard Wälchli, "The making of a festschrift, is it a ritual?," in *Methods in Contemporary Linguistics*, ed. Andrea Ender, Adrian Leemann and Bernahrd Wälchli (Berlin: Walter de Gruyter, 2012), 147–148.
51. S. Griswold Morley, "The Development of the Homage Volume," *Bulletin of the American Association of University Professors (1915–1955)* 15, no. 4 (1929): 293.

52. On the Humboldtian university reforms, see: Hans-Albrecht Koch, *Die Universität: Geschichte einer europäischen Institution* (Darmstadt: Wissenschaftliche Buchgesellschaft, 2008), chapter 5. Althoff's contributions will be discussed in more detail throughout Chap. 5.
53. Michael D. Gordin, *Scientific Babel: The Language of Science from the Fall of Latin to the Rise of English* (Chicago, IL: University of Chicago Press, 2015), 6.
54. Christophe Charle, "Patterns," in *A History of the University in Europe*, Vol. 3, ed. Walter Rüegg (Cambridge: Cambridge University Press, 2004). 33–80.
55. Owen Hannaway, "The German Model of Chemical Education in America: Ira Remsen at Johns Hopkins (1876–1913)," *Ambix* 23, no. 3 (1976): 145.
56. David Cahan, "Helmholtz and the shaping of the American physics elite in the Gilded Age," *Historical Studies in the Physical and Biological Sciences* 35, no 1 (2004): 2.
57. William Clark, "On the Dialectical Origins of the Research Seminar," *History of Science* 27, no. 2 (1989): 113.
58. Otto Kruse, "The Origins of Writing in the Disciplines: Traditions of Seminar Writing and the Humboldtian Ideal of the Research University," *Written Communication* 23, no. 3 (2006): 346–347.
59. Ibid., 345.
60. Lucy T. Benjamin Jr, Maureen Durkin, Michelle Link, Marilyn Vestal and Jill Acord, "Wundt's American Doctoral Students," *American Psychologist* 47, no. 2 (1992): 127.
61. An example is provided by the heated discussion about the measurement of higher mental processes in: Robert M. Ogden, "Oswald Külpe and the Würzburg School," *The American Journal of Psychology* 64, no. 1 (1951): 10–12.
62. Mitchell G. Ash, "Academic Politics in the History of Science: Experimental Psychology in Germany, 1879–1941," *Central European History* 13, no. 3 (1980): 278–282.
63. This was the so-called 'Tuberkulin affair'. See: Christoph Gradmann, "Robert Koch und das Tuberkulin—Anatomie eines Fehlschlags," *Deutsche Medizinische Wochenschrift* 124, no. 42 (1999): 1253–1256. See also Chap. 2, 35–36.
64. Florian Mildenberger, "Auf verlorenem Posten—der einsame Kampf des Heinrich Dreuw gegen Syphilis und Salvarsan," *Würzburger medizinhistorische Mitteilungen* 30 (2011): 171.
65. C. Meyer and S. Reiter, "Impfgegner und Impfskeptiker: Geschichte, Hintergründe, Thesen, Umgang," *Bundesgesundheitsblatt, Gesundheitsforschung, Gesundheitsschutz* 47, no. 12 (2004): 1183–1184.

66. Suzanne L. Marchand, *German Orientalism in the Age of Empire: Religion, Race, and Scholarship* (Cambridge: Cambridge University Press, 2009), xxxiii.
67. Rudi Paret, *Arabistik und Islamkunde an deutschen Universitäten: Deutsche Orientalisten seit Theodor Nöldeke* (Wiesbaden: Franz Steiner, 1966), 18.
68. Christiaan Engberts, "Gossiping about the Buddha of Göttingen: Heinrich Ewald as an unscholarly persona," *History of Humanities* 1, no. 2 (2016). See also Chap. 6, 184–185.
69. On the sometimes-tense relation between Nöldeke, Snouck, and Becker, see: Christiaan Engberts, "Orientalists at War: Personae and Partiality at the Outbreak of the First World War," in *Scholarly Personae in the History of Orientalism*, ed. Christiaan Engberts and Herman Paul (Brill: Leiden, 2019), 172–192.
70. For more on this project, see Chap. 2, 29–35.
71. For more on Nöldeke's (and Wundt's) reviews for the Literarische Centralblatt, see Chap. 4.
72. Wilhelm Wundt, *Erlebtes und Erkanntes* (Stuttgart: Alfred Kröner, 1920), 301. For more on the relation between Wundt and Fechner, see Chap. 1, pp. xx.
73. For more on Wundt's editorial work, see Chap. 3, 65–73.
74. For more on university appointments, see: Chap. 5.
75. The work of Althoff is particularly important to the argument made in Chap. 5.
76. The correspondence of Zarncke will be discussed at length throughout Chaps. 3 and 4.
77. David Shatz, *Peer Review: A Critical Inquiry* (Lanham, MD: Rowman & Littlefield, 2004), 111.

CHAPTER 2

A Helping Hand: Private Evaluation in Scholarly Correspondence

Letter Writing and the History of Science

The practice of writing letters has been central to the history of knowledge. Most long-term histories of European scholarship elaborate on a lengthy period in which the community of scholars could be characterized as a Republic of Letters. Discussions of this community often concentrate on the early modern period.[1] However, correspondences have been a salient feature of scholarly life both before and after this period. The cultural historian Peter Burke therefore argues that we should broaden our conception of the Republic of Letters and acknowledge that the practice of letter writing has shaped scholarship from the mid-fifteenth century to the modern day.[2] This observation does not imply, however, that this practice was essentially the same in the late nineteenth century as in the early modern age. Burke proposes to distinguish the early period of the Republic of Letters, lasting from about 1450 to 1850, from the subsequent period from 1850 to 1950. He characterizes the earlier period as "the age of the horse-drawn commonwealth" and the latter one as "the age of steam."[3]

The cut-off year of 1850 does not allude to any specific change in the practices of scholarly communication. However, the evolution of several developments, that can be traced back to the age of horsepower, had resulted in an age of steam in which scholarly correspondence had not so much lost its importance but rather acquired new meanings. In the early

© The Author(s), under exclusive license to Springer Nature Switzerland AG 2022

C. Engberts, *Scholarly Virtues in Nineteenth-Century Sciences and Humanities*, https://doi.org/10.1007/978-3-030-84566-7_2

modern period, one of the main functions of scholarly correspondence was the sharing of information: letters were arguably the most important channel for the circulation of scientific knowledge.[4] From the end of the seventeenth century onwards, scholarly journals like the *Journal des Sçavans* (1665) and the *Philosophical Transactions* (1665) would fulfil a similar function. In the light of this development, it has been argued that individual correspondences had lost most of their significance long before 1850.[5] At the same time, however, scholars continued to maintain extensive correspondences in a time in which international postal services were "becoming faster, cheaper and more reliable."[6]

Another consequence of the innovations of the nineteenth century was that it became increasingly easy to meet colleagues from faraway places. Of course, the learned societies that were founded in the seventeenth century, such as the *Académie Française* (1635), the *Leopoldina* (1652), and the Royal Society of London (1660), offered their members the opportunity to mingle. The nineteenth century, however, saw a boom of societies and congresses. One recent study confidently states that the ubiquity of international scientific congresses was a novel development of the nineteenth century.[7] Another study points at the establishment of the British Association for the Advancement of Science and other early Victorian national organizations as the impetus for the rapidly growing number of British scientific congresses.[8] A simultaneous and similar development occurred in the German states as well. One example is the *Deutsche Morgenländische Gesellschaft* (German Oriental Society), which was established in 1845 and which will play a central role in the next chapter.[9] These increased opportunities to meet other scholars in person did not, however, cause a decrease in scholarly correspondences.

Analyses of early modern scholarly letters emphasize their importance for creating and maintaining networks of people who had rarely—or even never—met in person. The establishment of mutual trust was therefore one of the primary objectives of most letters.[10] In this function, correspondences were an imperfect but necessary alternative to face-to-face meetings. The developments sketched above, however, raise a question about the function of scholarly correspondences in the late nineteenth century. If letters were no longer necessary to share knowledge or establish trust, what role did they play in the lives and careers of the scholars of this period? One answer to this question is that instead of establishing trust between faraway acquaintances, correspondences increasingly served to maintain cordial relations between scholars who occasionally met in person as well. One recent study argues that letters were ever more frequently used to discuss private issues.[11] Along similar lines, another recent study states that one of the most interesting characteristics of nineteenth-century scholarly correspondences is the way in which they functioned as a medium for confidential expression.[12]

While letters clearly did not lose their social functions, modern-day scholarship suggests that they retained a role in the exchange of knowledge as well. This role seems to have changed, however. If early modern correspondences largely functioned as precursors to print publishing, late nineteenth century correspondences were often used to exchange insights prior to the publication of books and articles. This has been observed by the editors of several recent volumes of late nineteenth and early twentieth-century scholarly correspondences. The introduction to the letters between the classical philologists Ulrich von Wilamowitz-Moellendorff and Eduard Norden mentions that the latter never "published anything which he had not first given to a friend to read."[13] The quantum physicist Wolfgang Pauli likewise refused to submit a ground-breaking article without first privately consulting his colleagues Niels Bohr and Werner Heisenberg.[14] The letters exchanged between the philosophers Alfred Schütz and Eric Voegelin provide another example. The editor of this correspondence argues that "both thinkers needed the other's participation and critique, and it is hardly possible to understand their works without taking their correspondence into account."[15]

Both the maintenance of friendship and the prepublication exchange of insights are common threads in the correspondences discussed in this chapter. Interestingly, the coexistence of both functions of letter writing invites exactly the kind of tensions that are central to this study as a whole: the complex interplay between the virtues of loyal collegiality and critical independence. All the protagonists of this chapter face questions and challenges that can be understood through this prism. To what extent does friendship allow for sharp mutual criticism? To what extent does friendship require honest and independent critique? To what extent does amicable and informal mutual criticism shape research and publishing? How do scholars deal with the oftentimes contradictory expectations of correspondents in their double capacity as both loyal friends and critical peers?

The first part of this chapter offers a detailed look at the correspondence of the orientalists Theodor Nöldeke and Michael Jan de Goeje, who had been good friends since their early twenties. In that capacity, they also continuously assessed each other's work. I will reflect on the interplay between their loyal friendship and critical judgements based on their extensive correspondence about De Goeje's edition of the *Annals* of the Persian historian al-Ṭabarī. Next, I will discuss a private collaborative relation that was not grounded in personal friendship: the paediatrician Otto Heubner's tentative early efforts to test the bacteriologist Emil Behring's first attempts to create a diphtheria blood serum. The final case study of this chapter is based on the correspondence between the psychologist and philosopher Wilhelm Wundt and his Leipzig collaborators. I will primarily be concerned with the questions why this correspondence contains fewer evaluations of to-be-published work than the letters of his orientalist and bacteriological peers and what this suggests about the potentially conflicting demands of loyalty and independence.

The Friendship between Nöldeke and De Goeje

Nöldeke and De Goeje belonged to the same generation: they were born within a half year of each other in 1836. Nöldeke was born in Harburg, a small town close to Hamburg.[16] In 1849, the family moved to Lingen, a village close to the Dutch border. At a young age, his father, a high school teacher, taught him Latin, Greek, and the basics of French and English. At the age of fifteen, he read Gesenius' *Hebräische Grammatik* and soon his Hebrew was better than that of his teacher. In 1853, he began his studies in Oriental languages in Göttingen with the leading Old Testament scholar Heinrich Ewald, whom his father knew from his own student days. Ewald was known as a demanding and uncompromising man and his teaching was unstructured.[17] He could, however, also be an inspiring educator and managed to challenge his acquaintance's son to live up to his high expectations. In 1856, Nöldeke obtained his doctorate on the basis of an essay about the history of the Quran.[18] The next year he travelled to the Netherlands to study the *Legatum Warnerianum*, the collection of Oriental manuscript of Leiden University. This is where he met Michael Jan de Goeje.

De Goeje was born in the Frisian village of Dronrijp as the son of a Protestant minister.[19] His father taught him classical and modern languages. His father died when de Goeje was eighteen, but his family raised money to send him to Leiden to study theology. As a future theologian, he had to familiarize himself with some Semitic languages, especially Hebrew. He enjoyed these studies so much that he decided to quit theology and become a Semitist instead. T.W.J. Juynboll, professor of Eastern languages, referred him to the renowned Arabist and professor of Modern History, Reinhart Dozy. Dozy proved to be as inspiring a teacher as Ewald and would remain an example for De Goeje throughout his career: "To collect and critically rework Arabic texts that had to serve as sources of a certain part of the history of civilisation, and then publish the results from these studies in a tasteful fashion, just like Dozy [...] became and continued to be his scholarly ideal."[20] Juynboll awarded him his doctorate for an Arabic text edition and a Latin translation of excerpts of an Arabic geographical work, al-Yaqubi's *Kitab al-Buldan* in 1860. His academic career therefore took off somewhat slower than Nöldeke's, who arrived in Leiden as a freshly minted doctor in 1857.

De Goeje was duly impressed by the erudition of his German peer. Half a century later, he still recounted the strong impression that Nöldeke made on him: "Every day I felt the contrast between your brightness, your ingenuity, your maturity, and my own schoolboy daftness, my slow thinking, my clumsiness."[21] Notwithstanding De Goeje's self-perceived daftness, the two men got along very well. The De Goeje family even functioned as a foster family for Nöldeke. More than sixty years later, he would still reminisce about Leiden family life in a letter to De Goeje's former student Christiaan Snouck Hurgronje: "With a certain reverence I still remember the evenings spent with De Goeje's mother, who lived in very modest conditions, and who had been born in Lingen, by the way."[22] The friendship between the two scholars would persist until De Goeje's death in 1909.

Their time together in Leiden ended earlier than expected. In 1858, Nöldeke left Leiden to rewrite his dissertation for a competition sponsored by the Parisian *Académie des Inscriptions*, on the basis of manuscripts in the Berlin Royal Library. He left Leiden reluctantly, but his hard work paid off. The next year, he was the youngest of the three winners of the competition, the others being Aloys Sprenger and Michele Amari. The German translation of this work, the "first masterpiece of his career," was published in 1860, under the title *Geschichte des Qorāns*.[23] That same year Nöldeke moved to Göttingen, first as a librarian and, after finishing his *Habilitation*, as a *Privatdozent* in Semitic languages. In 1864, he accepted a position as *Extraordinarius* in Kiel, which would be made into a full professorship four years later. In 1874, he was appointed to the Chair of Semitic Languages at the newly established *Reichsuniversität Strassburg*. Although other prestigious universities, such as Vienna, Berlin, and Leipzig, tried to entice him, he would stay in Strasbourg for the remainder of his career.[24] He retired in 1906, at the age of seventy, but continued to publish for twenty more years. He died in 1930 at the age of ninety-four.

The career of his friend De Goeje was equally successful. Before he had finished his doctorate, he was already appointed assistant curator of the *Legatum Warnerianum*. Because this seemed to be a dead-end job, however, he started looking for jobs outside academia. In 1864, he wrote to Nöldeke that he might have to give up his scholarly ambitions.[25] In 1866, Leiden University finally offered him an assistant professorship in Eastern languages. Three years later, he was appointed as *Interpres Legati Warneriani*, head curator of the manuscript collection, as well as full professor. He would hold on to this professorship until his retirement in 1906. Unlike Nöldeke, who published on a myriad of Semitic languages, De Goeje was primarily interested in one language: Arabic. Most of his publications were text editions. From 1870 onward, he published the *Bibliotheca geographorum Arabicorum*, an eight-volume series of works by Arabic geographers. His most impressive achievement was his edition of the *Annals* of al-Ṭabarī, published between 1879 and 1901.

Between the first meeting of Nöldeke and De Goeje in 1867 and the latter's death in 1909, more than half a century had passed. In 1907, Nöldeke fondly looked back at half a century of friendship: "These days, my dear, 50 years must have passed since we first met. There was an immediate connection between us, even if we did not yet know how close and lasting our friendship would become. [...] I still often think with exceptional fondness of your lovely mother and her cosy home. It is a long, long time ago, but it often feels as if it was only yesterday."[26] In the preceding half century, they had not only shared scholarly insight but intimate details about their privates lives as well. De Goeje, for example, described his love life in glowing terms:

> I would love to introduce you to a couple of happy people, who cannot think about sad things, for whom war does not exist, who cannot even be saddened by the cholera. One of the two is probably completely unknown to you. It is a charming, lovely girl, with sensible and sweet eyes, with beautiful hair, with a musical voice. [...] The other might be better known to you, maybe you even have his portrait in your album. [...] he is as happy as he never dreamed, he would ever be. Indeed, this happy young man is none other than your old friend, Jan de Goeje. At the moment our happiness is like an oasis in the desert. The cholera besieges our city, and all the people are dreary and sad. [...] And now people throng together from all directions

to lavish themselves with the sight of a few blissful people, who do not know gloom and dreariness and who are not capable to believe in sadness.[27]

He would marry this girl, Wilhelmina Leembruggen, the next year.

Nöldeke and De Goeje did not only share joyful life events. When the latter retired, he shared his worries about his imminent mental and physical decline: "Yes, the time of retirement is coming into view. How I think about it now, I hope that life will draw to an end as well. I do at least hope that I will not have to subsist as a caricature of myself. I wish you the same, my old loyal friend"[28] They also discussed the lives of mutual friends, such as the Cambridge orientalist William Wright: "If Wright […] would die! Terrible! Then *both of us* will have to promise each other to stay alive for a really long time"[29] De Goeje also adorned his study with framed portraits of both Nöldeke and Wright.[30] Meanwhile, there was room to discuss family hobbies as well: Nöldeke sent German stamps to Leiden for the collection of De Goeje's younger brother. De Goeje sent Nöldeke's family stamps from the Netherlands and the Dutch Indies.[31]

Most of the correspondence between Nöldeke and De Goeje, however, was about their scholarly endeavours: they wrote long letters in which they scrutinized each other's work. It took some years to settle on an appropriate balance between private and public criticism. In the mid-1860s, for example, Goeje was unhappy to find out that Nöldeke's privately shared critique of his *Historia khalifatus Omari II, Jazídi II et Hischámi* could also been found in a book review in the *Göttingische Gelehrte Anzeigen*.[32] He insisted that such observations should only be shared in private. From this moment onward, Nöldeke would only comment on his friend's work in his letters. His most time-consuming evaluative endeavour would be his proofreading of long sections of the al-Ṭabarī edition.

Discussing the *Annals* of al-Ṭabarī

The History of Prophets and Kings by Muhammad ibn Jarir al-Ṭabarī, commonly referred to as the *Annals* of al-Ṭabarī, was a revolutionary work of historiography. It was completed in the tenth century, and it was the first universal history that covered the whole period between the creation and the author's own time.[33] From the seventeenth century onwards, European scholars had published excerpts of the *Annals*.[34] The most comprehensive edition available during De Goeje's student days was Johann Ludwig Kosegarten's three-volume edition of a Berlin al-Ṭabarī manuscript.[35] The fact that De Goeje's edition would eventually consist of three series of books instead of three volumes shows just how daunting a task he had undertaken. The first series counted six books, the second one three, and the fourth series consisted of four books. In total they contained more than 8000 pages. De Goeje also assembled two additional volumes with indexes, *addenda et emandanda*, and a glossary.

Kosegarten's edition had been based on only one manuscript that was neither of a particularly high quality nor remarkably authentic. At the time, such editions seemed to be the best that could be achieved because no complete copies of the *Annals* were known. In 1858, De Goeje first discussed the possibility to nevertheless publish a complete edition.[36] Shortly after obtaining his doctorate, he published a note in the *Zeitschrift der Deutschen Morgenländischen Gesellschaft*, in which he discussed some excerpts that he had discovered in the Bodleian Library. He argued that this Oxford manuscript would "be of great service to the future editor of Tabari."[37] Almost three years later, he expressed the ambition of undertaking this task himself for the first time: "it is a shame that Tabari is still unpublished. Maybe I will take it upon me one day."[38] At this moment, the endeavour still seemed to be too ambitious. After all, he was only a twenty-nine-year-old assistant curator without any assurance that he would eventually succeed in pursuing an academic career.

Though De Goeje never lost his interest in the *Annals*, he did not seriously plan an edition of the full work before December 1872. That month he received a letter from Albert Socin, an orientalist from Basel. Socin's former teacher, the theologian Johann Jakob Stähelin, had announced that he was willing to spend "a considerable sum" on the publication of the *Annals*.[39] Socin could not undertake this project himself, but his friend Otto Loth had informed him that De Goeje had a vivid interest in the work. It was not difficult to enthuse the latter and in early 1873 he eagerly discussed his plans with some colleagues.[40] The next year, the first outlines of what was to become an international al-Ṭabarī consortium started to take shape. Loth enthusiastically joined the enterprise, while Nöldeke more reluctantly agreed to edit a section.[41] Nöldeke's former student Eduard Sachau and the German orientalist and diplomat Andreas David Mordtmann started looking for manuscripts in Constantinople.[42] Throughout the next few years, a large number of German, Dutch, French, Austrian, Italian, and other scholars would join the consortium.[43]

Notwithstanding his initial reluctance, Nöldeke ended up spending a lot of effort on the project. He did not only edit the *Annals* section about Sassanid history, but he also published a German translation of the text.[44] During the more than twenty years it took to finish the endeavour, Nöldeke contributed in other ways as well. At its beginning, he was involved in drawing up the editorial guidelines.[45] Since there was no universally accepted template for publishing Arabic texts, De Goeje had to create his own guidelines on issues like the use of diacritical points, the way to refer to Quranic and other quoted and paraphrased texts, the additional information to be provided in footnotes, and the criteria for the collection of words and phrases for the glossary. Though De Goeje accepted most of Nöldeke's suggestions, he did not adopt them all. Nöldeke was, for example, strongly opposed to asking the editors to collect words and phrases for the glossary. He argued that this would be impossible for him because he did not have access to the necessary Arabic lexicons. De Goeje nonetheless asked all editors to record proper names, place names, and proverbs. He did, however, mitigate his request for the collection of words that had not been recorded previously. He asked his collaborators to collect these only if this would not cause too much trouble and postponed his final decision about the form of the indexes and glossary.[46]

The Minutiae of Arabic

Of course, Nöldeke's involvement with the actual editorial work was much more time-consuming than his contributions to the editorial guidelines. He did not only edit hundreds of pages himself, but he also served as an unofficial, unpaid proofreader for texts edited by others. De Goeje's al-Ṭabarī correspondence contains Nöldeke's extensive commentary on the contributions of, among others, Ignazio Guidi and Pieter de Jong.[47] Their work was evaluated in detail by both Nöldeke and De Goeje. Nöldeke later recounted how closely De Goeje was involved with the work of all editors: "[During the making of al-Ṭabarī], De Goeje, who indeed is the foremost living Arabist, supervised every single thing, and because four eyes always see more than two, much was improved of even the best collaborators."[48] Following this line of reasoning, it is self-evident that even a meticulous editor like De Goeje could benefit from a second pair of eyes. It was clear from the outset that Nöldeke would take on this task. He closely examined hundreds, if not thousands, of pages of the *Annals*, that

De Goeje had edited himself. He formulated his commentary in the way that his friend liked best: long lists of detailed comments.[49]

These lists all shared the same format. At the top of each list, Nöldeke would indicate the part of the series on which he commented. The rest of such a list would show page and line numbers in the left margin, with the related commentary to the right. Sometimes, he only corrected a minor printing error. Often, however, his comments merited extensive elaboration because most manuscripts were not as clear as the editors may have wished. Sometimes this was due to the bad quality of the available copies, at other times it was caused by the idiosyncrasies of the Arabic script. Most manuscripts lacked vowel points and other diacritical marks. These were often absent, even in the manuscripts that De Goeje and Nöldeke considered to be of high quality, such as an *Annals* fragment kept in Leiden, which Nöldeke described as "a copy of a *very good* codex made by a completely ignorant copier, which, however, has only few diacritical points and vocals."[50] Yet, these points and vocals would have been very useful, especially because all the *Annals* manuscripts contained a lot of hitherto unknown words and proper names.

In some cases, the challenge was not only to reconstruct the original text without the help of diacritical marks, but also to judge the authenticity of the diacritical marks in the manuscripts. In the same letter as the one quoted above, Nöldeke complained that it was "inconvenient that various later hands have added diacritical marks to the manuscript, which cannot always easily be distinguished from the original hand." The lack of diacritical marks and reasonable doubts about their authenticity left ample room for doubts about the correct transcription. Therefore, one recurring point of discussion between Nöldeke and De Goeje concerned the minutiae of the Arabic language, such as the requirement to either delete or add diacritical marks like the *sukūn* and the *tanwīn*.[51]

The most common way to resolve vagueness and ambiguity in manuscripts was comparison with other texts. The *Annals* contain so many quotations, that part of al-Ṭabarī's work is best described as that of a compiler or editor.[52] Some of his material is drawn from earlier historiographers.[53] In other places, he extensively quotes ancient poetry, which often contained historical narrative, as well.[54] Other sources that the editors could fall back on were the works of Arabic scholars who quoted from the *Annals*. Nöldeke could, for example, finish parts of his work without having al-Ṭabarī's full text at his disposal because these excerpts had been incorporated into the work of ʿAlī Ibn al-Athīr.[55] He also made use of other Arabic chronicles and geographical works, some of them preceding al-Ṭabarī, such as al-Balādhurī's works, others partially based on the *Annals*, such as the works of Ibn Khaldūn and al-Masʿūdī.[56] Nöldeke proposed further changes to al-Ṭabarī's postscript on the basis of his reading of Ibn Hisham and recommended corrections of Guidi's text after a comparison with the works of Yāqūt and Ibn Zubayr.[57] Probably the best-known source of the *Annals* was the Quran; doubts about the meaning and orthography of words that could also be found there were easily resolved.[58]

Not all Nöldeke's proposed corrections were based on comparisons between al-Ṭabarī's text and those of others. Often, his comments were rather based on a general understanding of Arabic writing styles. Such assessments had to account for the fact that al-Ṭabarī had used the work of many authors from a wide variety of genres. The *Annals* replicate short single sentence reports and medium-sized reports of a few dozen lines, as well as more extensive longer reports.[59] A lot of these are written in prose, but the work also contains vast quantities of historical poetry.[60] All these sections are written in the literary style that is characteristic of that specific genre. Ancient pre-Islamic epics and odes, for instance, are known for their highly stylized language: "The number and complexity of the measures which they use, their established laws of quantity and rhyme, and the uniform manner in which they introduce the subject of their poems […] all point to a long previous study and cultivation of the art of expression and the capacities of their language."[61]

Even when the aesthetic demands of poetry did not suggest any particular reading, Nöldeke sometimes rejected certain suggestions because they were simply "barbaric" in his eyes and could therefore not be authentic.[62] Other words and constructions were rejected by Nöldeke because they were too stilted to suit the plain prose texts in which they appeared.[63] Sometimes he suggested that certain words and constructions had to be changed, not because they were stylistically improper, but because they were uncommon, and a rather obvious alternative was available.[64] If this obvious alternative clearly differed from the manuscript, he assumed that the manuscript was less trustworthy than his own critical judgement: "How limited is the authority of codices in these matters!"[65] However, even with his well-developed sense of language, Nöldeke had to admit that some excerpts remained incomprehensible.[66] This is hardly surprising, since much of the poetry in the *Annals* still conjures up varying interpretations among modern-day scholars.[67]

Nöldeke's attempts to assess the poetic parts of the *Annals* were further aided by rhyme and metre. He corrected De Goeje several times on the basis of rhyme schemes.[68] He more often referred to metre, however, because this was one of his friend's main weaknesses. Shortly after finishing his studies, De Goeje already acknowledged the insufficiency of Dozy's teaching on this topic: "Dozy read a lot with us, but teaching grammar was not to his taste. By now I have learned it the hard way, but not sufficiently yet. I will amend it, however, just like my knowledge of metrics, which could have been more comprehensive."[69] At the end of his career, it was still one of his least favourite subjects: "I hate didactic poetry and books about metrics."[70] Looking back at his friend's career, Nöldeke had nothing but praise for his grammatical advances, but he also noted that De Goeje had never fully caught up on metrics: "[I]t should be highly respected, how De Goeje later found his way around grammatical refinement. Only in one respect he never became confident, in metrics, because he apparently never had an ear for music at all. Even in his last text editions [...] some disruptions of the metre occur."[71]

This last assessment was made nine years after De Goeje's death in a letter to his student and successor Christiaan Snouck Hurgronje. But even during De Goeje's lifetime, Nöldeke shared this criticism with some trusted colleagues. To the Budapest orientalist Ignaz Goldziher, he wrote: "I have given [De Goeje] some text corrections, some of which he rightfully showed me to be false. It is strange, however, that De Goeje has so little sense of rhythm that violations of the metre still happen to him."[72] Therefore, it is not surprising that remarks about metre are a recurring theme in Nöldeke's comments on excerpts of the *Annals* that had been edited or approved by De Goeje.[73] Nöldeke's lists of comments often contained added remarks by De Goeje, as well. But, although these remarks show that he did not accept all Nöldeke's corrections, such disagreement is not displayed in his handwritten comments in reaction to any metric proposal.

A quick look at Nöldeke's comments shows the extent to which his private evaluation shaped his colleagues' end product. The scholarly end product of most orientalists' endeavours was a text, whether it was a grammar, a chrestomathy, a textbook, or a text edition. These texts lent themselves pre-eminently to precise and exhaustive evaluation. The knowledge of colleagues could easily and immediately be incorporated. Therefore, a critically independent friend was a valuable asset. Because loyalty could be expressed exactly by sharing critical assessments of each other's work, the potential tensions between loyal collegiality and critical independence could be resolved with relative ease among friends who were willing and able to profit from each other's critical insights. However, as the next section shows, this way of dealing with the simultaneous demands of loyalty and independence was not as self-evidently applicable to other disciplines that were less centred on the individual scholar's engagement with texts.

Clinical Testing as Evaluative Practice

The private correspondences of nineteenth-century bacteriologists do not show the same level of constant mutual evaluation of to-be-published texts as those of their orientalist peers. This does not imply, however, that they were not interested in assessments of their work before it would be made public. In fact, they had even better reason to have their findings meticulously checked than the orientalists. After all, these findings were often not texts, but new drugs and treatment regimens that could cure or kill people. Then, as now, one of the most important steps in assuring the efficacy and safety of these new cures and regimens was to test them not only on animals but on people as well. Such tests are nowadays known as clinical trials. Historical overviews of the development of this practice often start with James Lind's testing of potential cures for scurvy in 1747.[74] By the end of the nineteenth century, however, there were still no generally accepted guidelines and requirements for clinical testing. The German medical community was reminded of its importance, however, when Robert Koch, one of the country's most respected bacteriologists, became embroiled in an affair that threatened the credibility of the discipline.

In 1890, the tenth International Medical Congress met in Berlin. The highlight of this meeting was Koch's announcement that he had found a cure for tuberculosis, one of the most widespread and deadly diseases of the time. Later that year, he published his findings about the active substance, which he had named *Tuberkulin*, in the *Deutsche Medizinische Wochenschrift*.[75] The news was enthusiastically received: German newspapers reported Koch's accomplishment with patriotic pride and tuberculosis patients from all over the country gathered in Berlin in the hope to benefit from the new cure.[76] But the enthusiasm wore off soon. The pathologist Rudolf Virchow argued that *Tuberkulin* actually accelerated the pathological process and the clinician Ottomar Rosenbach demonstrated that it could have dangerous side effects.[77] The fact that Koch could neither produce the guinea pigs that he had supposedly cured nor the exact composition of *Tuberkulin* left him vulnerable to their accusations.[78] The Prussian state authorities, who supported Koch, also apprehensively emphasized the unfinished nature of the *Tuberkulin* research programme in their internal communication.[79]

In the end, *Tuberkulin* failed to live up to the expectations of the state, the press, tuberculosis patients, and the medical community. Koch would be the only beneficiary of the whole episode: the initial enthusiasm had enabled him to secure state support for his own research institute, the *Institut für Infektionskrankheiten* (Institute for Infectious Diseases). The reputation of the relatively young discipline of bacteriology, however, was badly tarnished. Therefore, the first people to make a new, potentially revolutionary, discovery in this field knew that they would face an uphill battle to convince both their peers and the public at large of the credibility of their findings.

The first new major bacteriological discovery would be a blood serum that could be used both as a cure and as a prophylactic against diphtheria. Its developers, Emil Behring and Erich Wernicke, both worked as research assistants at the Institute for Infectious Diseases. Behring was born in 1854 in Hansdorf, a village in modern-day Poland. After receiving his medical doctorate at the University of Berlin in 1878 and his license to practice medicine in 1880, he worked as an army doctor in eastern Prussia.[80] Between 1887 and 1889, he was stationed at Karl Binz's Pharmacological Institute in Bonn, after which he was sent to the Hygienic Institute at the University of Berlin, where he worked under Koch. When Koch moved to the Institute for Infectious Diseases in 1891, Behring was appointed as one of his assistants. At the Hygienic Institute he met Erich Wernicke. Wernicke was five years his junior and was an army doctor too.[81] In 1891, he was also appointed at the Institute for Infectious Diseases, where he found himself "caught by Behring's towering idiosyncratic character."[82]

Under Behring's leadership, Wernicke worked on the diphtheria serum. His help was desperately needed because Behring's poor health prevented him from performing animal trials. Wernicke and his fiancée therefore took care of most of these tests.[83] Even though they were promising, Behring realized that they might not convince their medical peers of the efficacy of their serum: the unsatisfactory outcomes of the still recent *Tuberkulin* affair threatened the reception of their findings. In a letter to a sympathetic paediatrician, he mentioned that he hoped "to make use of the experiences of *Koch* in the tuberculosis treatment and to be spared similar setbacks."[84] He finished this letter with the statement that he would "*[...] rather wait some more years with further publications, than present something doubtful now.*" Looking back on these days, Wernicke also underlined the importance of Koch's fiasco. He recounted how "a major medical authority" dismissed their findings with a condescending comment: "The serum is a slippery substance, on which its discoverers will slip." Other commentators even explicitly referred to the *Tuberkulin* affair and characterized the diphtheria serum as a "similar bacteriological scam."[85]

It was, therefore, obvious to Behring and Wernicke that it would be hard to convince their peers of the merit of their discovery. Behring even feared that decisive scientific proof might not be enough to placate the sceptics. He cynically wrote in his diary that it was important to "work on the emotions, not on reason, when one wants to carry away the crowd."[86] Still, both men knew that reason should not be overlooked: they needed compelling proof of the efficacy of their serum. Carefully conducted clinical trials by trustworthy colleagues seemed to be the most promising way to work on both the emotions and reason. The first tests were carried out in the clinic of the Berlin paediatrician Ernst von Bergmann, in December 1891.[87] These were small-scale trials with largely unsatisfying results. The first clinical tests on a large enough scale to yield significant results started one year later under the supervision of the paediatrician Otto Heubner. Behring and Heubner would develop a relationship based on mutual respect and shared aims rather than on friendship.

Testing the Diphtheria Blood Serum

Otto Heubner was born in 1843 in a village in the south of Saxony.[88] Even though he was trained as an internist, the large number of children visiting the District Policlinic in Leipzig, where he worked from 1876 onwards, pushed him towards paediatrics. In 1886, he was appointed as professor of Paediatrics in Leipzig and, two years later, he founded a children's hospital with the donations of wealthy patients and other sponsors. Since diphtheria mostly effects children and Heubner's management of the hospital was widely praised by his contemporaries, he was both a highly motivated and a very suitable collaborator on the first large-scale clinical trials of the new serum. He was not the only clinician testing the serum after Bergmann's earlier disappointing early trials. In one of his letters, Behring does refer to some trials under the leadership of the Berlin paediatrician Eduard Heinrich Henoch.[89] Heubner, however, made most of the observations of the effect that the early versions of the serum had on people.[90] Especially throughout 1892, Behring flooded him with requests to test new versions of his serum and to answer a myriad of questions about their effects.

One question that Behring repeatedly asked was if Heubner could determine whether the antiserum provided a specific cure against diphtheria.[91] Even if the test results on animals strongly supported this conclusion, he was disappointed to learn that results of the clinical tests were not as straightforward.[92] One reason for this could be that some of the tested children suffered from other diseases, as well; diphtheria was often accompanied by a streptococcus infection.[93] Another reason was that the serum could have different effects on different groups of patients. Henoch, for example, had not tested the serum on seemingly mild or beginning cases of diphtheria.[94] The fact that a relatively large number of the people he had tested—who all had a negative prognosis to begin with—were not cured, did not necessarily prove anything about the efficacy of the serum on milder and more recent infections. Behring therefore asked Heubner to test the serum on children with mild, moderate, and severe cases of diphtheria and pressured him to make a clear distinction between his findings about all three categories in his report.[95]

It was not enough for Behring to know whether his serum was a specific cure for diphtheria. He also asked Heubner to establish the appropriate dosage. Because it was unlikely that there would be one dose that would cure diphtheria in both its earlier and later stages, he repeatedly asked to look for both the "curative minimal dosage" and what increase in this minimal dosage would be effective in fighting the more advanced stages of the disease.[96] Behring was also interested in the serum's side effects.[97] Heubner's findings about this topic seemed to be ambiguous: in the spring of 1893, however, Behring happily concluded that the serum was "absolutely safe" for human use.[98] Finally, Behring asked for statistical data about every circumstance that could be relevant in determining just how effective the serum was. Even if the tests did not include control groups, as required in most modern clinical trials, Behring asked Heubner to also collect data on children that he had treated for diphtheria before testing the blood serum.[99]

At the end of 1892, however, the results from Heubner's tests were still not decisive. Although they suggested that the serum was safe, they had not provided clear indications for the optimal dosage. At this moment, Behring could still not rule out the possibility that the serum was not effective at all! Luckily, the provisional results from Heubner's testing were promising enough for Behring and Wernicke to find outside support. The *Höchster Farbwerke*, one of Germany's major chemical manufacturers, showed an interest in producing the serum. The *Farbwerke* committed themselves to funding large-scale additional research in 1893. If the results of this research would be encouraging, they promised further investments in the serum's development and marketing.[100] At the same time, Paul Ehrlich, another member of the Institute for Infectious Diseases, teamed up with Behring and Wernicke to investigate new ways to determine the effectiveness and recommended dosages of the serum.[101] In his earlier research, he had primarily focused on the quantitative relationship between serum and antibodies.[102] On the basis of this expertise, he was able to improve on Behring's and Wernicke's earlier efforts to establish the appropriate serum dosage. While he was reluctant to give him too much credit, twenty-five years later, Wernicke admitted that "nobody will question [Ehrlich's] epochal genius [...] when it comes to establishing the impact of serums."[103]

The combination of findings from new clinical trials in Berlin and the efforts of Paul Ehrlich convinced August Laubenheimer, member of the board of the *Farbwerken*, to deliver a positive verdict on the serum's efficacy and commercial viability.[104] The final affirmation of its commercial viability was the result of a discussion of the trials at the Imperial Health Office in Berlin. About fifteen medical doctors, among whom Behring, Ehrlich and Koch, convened at a meeting chaired by the director of the office, Dr Karl Köhler, and attended by *Ministerialdirektor* Friedrich Althoff. After all the attendees had vowed for the efficacy and safety of the serum, it was decided that it could be sold at pharmacies as a prescription drug.[105] From this moment onwards, the *Hoechster Farbwerke* acted quickly on Laubenheimer's advice from earlier that year. Already in November 1894, the serum entered the market. The festive opening of a brand-new serum production facility was attended by, among others, Behring, Ehrlich, Koch, Köhler, and Althoff.[106]

Evaluating Texts Versus Evaluation Serums

A comparison between Heubner's evaluation of Behring's serum and Nöldeke's evaluation of De Goeje's texts allows for a better understanding of the characteristics—or even idiosyncrasies—of both evaluative processes. The processes share some striking features. Both Nöldeke and Heubner dealt intensively with the work of a colleague over a long period. Both were tasked with pinpointing the shortcomings and confirming the strengths of the results of many years of research before they would be shared with the public at large. Both worked for free without expecting material personal benefits. The differences between their efforts, however, are as striking as the similarities. It could, for example, be argued that their motivations were very different. In addition, I would argue that there are at least three other major interrelated differences.

The first difference is the fact that the objects of evaluation are very dissimilar. De Goeje and his collaborators produced thousands of pages of text and Nöldeke evaluated them closely. Behring and his collaborators produced reports of their findings, too, and these texts were duly published in relevant journals or as independent volumes.[107] They did not, however, ask Heubner or Ehrlich to carefully read any of these texts. Instead, Behring asked Heubner repeatedly to evaluate the efficacy and side effects of new versions of his serum. Likewise, he asked Ehrlich only to help him to find ways to produce the serum more efficiently and to establish the strength of individual batches.[108] The emphasis on the evaluation of a material instead of a textual object called for an effort that was very different from the task to which Nöldeke had committed himself.

The second difference between Nöldeke's and Heubner evaluations has to do with the extent to which they repeated the intellectual process through which the creator of the original product had gone. Nöldeke's evaluative efforts consisted of continually asking questions that were the same as those of De Goeje and his collaborators: "Is this sentence grammatically correct?" "Do the metre and rhyme in these lines follow an established pattern?" "Could these words be the same as differently spelled words in other texts?" Therefore Nöldeke's intellectual effort was largely a reiteration of the efforts of the editors. Heubner, on the other hand, was not asked to repeat Behring's thought processes, at all. If Nöldeke was invited as a commentator because he was deemed to possess the same type of expertise as those whose work he reviewed, Heubner's evaluation was important because he possessed other expertise and resources. His task was not to assess the process by which the serum had been made, but to find out if it did what its creators hoped it would do. His added skills and resources were a long experience in observing and diagnosing diphtheria in people, the ability to manage large-scale clinical operations, and access to a statistically significant number of diphtheria patients.

Thirdly, there are major differences in the work environment that shaped both evaluative efforts. Text editing is a largely solitary endeavour. De Goeje's collaborators worked alone in their offices. They sent their work to the editor-in-chief, who would proofread it in the solitary confines of his own study. De Goeje would then send these manuscripts to Nöldeke, who would also spend long days alone with them. It is exactly the solitary character of this work that makes it possible for reviewers and editors to reiterate the intellectual processes of those who came before them. Behring's serum, however, was created in a laboratory setting, in close personal collaboration with colleagues such as Wernicke and Ehrlich, and with the help of a small army of paid and unpaid assistants.[109] Because this collaborative effort could not be completely reiterated, it was self-evident that its evaluation would focus on the serum's efficacy rather than on the process of its creation.

Nöldeke and Heubner were therefore both encouraged to exhibit critical independence for different reasons. The solitary character and textual emphasis of the orientalists' work encouraged a specific way to engage with a text. The evaluator did not have to accept it at face value but was largely able to independently judge all its features on a case-by-case basis. Critical independence was encouraged because a highly detailed assessment was feasible. A similar type of detailed assessment was not possible for Heubner because of the collaborative and non-textual character of the diphtheria serum. The potentially far-reaching consequences of contributing to the approval of an ineffective or even harmful blood serum, however, provided a powerful incentive to assess its impact critically, especially in the light of the widely publicized *Tuberkulin* fiasco.

In both cases, no major conflict between critical independence and collegial loyalty occurred. The lack of tension in the relationship between Behring and Heubner can largely be attributed to the fact that the former realized that the latter's critical independence was essential to the eventual acceptance of his serum. He was even willing to postpone the public presentation of his findings for years if Heubner's initial tests would have been unsatisfying.[110] The apparent ease with which Nöldeke and De Goeje accepted each other's critical scrutiny is grounded in their close relationship. Precisely because they trusted each other as friends, they were able to gratefully accept and benefit from each other's independent criticism.

Working with Wundt

Even though there were different ways in which collegial loyalty and critical independence could be reconciled in private correspondences, the example of Wilhelm Wundt shows that it was not self-evident that scholars would be able to balance both expectations in such a fruitful way as Nöldeke, De Goeje, Behring, and Heubner. Wundt was born in Neckerau, just outside of Mannheim.[111] When he was four years old, his parents moved to the small town of Heidelsheim. During his school years, he did not stand out; in primary school, he was an absent-minded daydreamer and his high school teachers at the *Gymnasium* thought he was dull-witted and lazy. Only after he was sent to the Lyceum in Heidelberg did he prove himself to be an above-average student. He spent his first year as a student enjoying cultural life in Tübingen, after which he returned to Heidelberg to pursue his medical degree with more vigour.[112] After completing his medical studies, Wundt held an assistantship in Heidelberg with Ewald Hasse. After a short stay in the physiological laboratories of Johannes Müller and Emil Du-Bois Reymond in Berlin, he returned to Heidelberg to become the assistant of Hermann von Helmholtz. Shortly after leaving his laboratory in 1863, Wundt was appointed as *Extraordinarius* at his alma mater. This job did not provide him with a regular salary but was able to supplement his income by writing textbooks.[113]

During these years, his background in physiology and growing interest in philosophy started to converge. He increasingly turned his attention towards the cutting edge of the physiology of perception and the philosophy of mind. His future academic career was still very uncertain, however. In 1872, he wrote to his fiancée, Sophie Mau: "My actual scientific pursuits, specifically those that are done for science's sake and not just to make a living, mostly occupy a fringe area between physiology and philosophy, suspect among respectable scientists, where not much prestige can be gained at the moment."[114] However, his 1874 book *Grundzüge der Physiologischen Psychologie* (Principles of Physiological Psychology) was a well-received exposition of his approach.[115] In the same year, he was appointed to the Chair of Inductive Philosophy at the University of Zürich.

One year after his arrival in Switzerland, he was called to Leipzig University. He was appointed at the Faculty of Philosophy to study the relationship between philosophy and the natural sciences. Here, he met two of the founding fathers of physiological psychology: Ernst Heinrich Weber and Gustav Theodor Fechner. Wundt and Fechner soon developed a cordial relationship. Shortly after his arrival in Leipzig, Wundt established what is now seen as the first laboratory for experimental psychology, the *Institut für experimentelle Psychologie* (Institute for Experimental Psychology). From the autumn of 1879 onwards, a growing number of students, such as Friedrich Tischer, Emil Kraepelin, G. Stanley Hall, and

James McKeen Cattell, worked on experimental projects under Wundt's supervision. Wundt later argued that the *de facto* establishment of the laboratory can be traced back to this autumn.[116] Though modern-day scholars disagree about the extent to which these developments can be perceived as the birth of psychology as an independent discipline, most agree about the importance of this laboratory to the history of experimental psychological research.[117]

Though Wundt exchanged letters with many people, his correspondence contains very few in-depth discussions with outsiders of the work done in his laboratory. His correspondence with Fechner is the main exception. Between 1879 and Fechner's death in 1887, a lively exchange of ideas took place between them. Their earliest preserved correspondence is about an issue on which they strongly disagreed, Henry Slade's spiritual abilities. Slade was a self-proclaimed medium, who arrived in Leipzig in 1877, after his claims to be able to move objects by channelling unknown forces had been debunked in New York and London. He nevertheless caught the attention of the Leipzig astrophysicist Friedrich Zöllner, who organized sessions with Slade at his house.[118] Wundt and Fechner both attended at least one of these sessions.[119] Wundt did not believe in the veracity of Slade's performance and attacked his trustworthiness in a scathing booklet.[120] Fechner disagreed with Wundt's harsh judgement and defended Slade in a long private letter.[121] After a short exchange of arguments, they agreed to disagree. "Why would we keep on arguing, because I would rather not quarrel with you about this issue, now that we have convinced each other that we cannot lecture each other about those things about which we disagree," Fechner wrote.[122]

With the Slade discussion out of the way, the correspondence between Wundt and Fechner would mostly focus on the main legacy of Ernst Heinrich Weber, a principle known as Weber's law. It states that "the increase in any stimulus necessary to make a noticeable difference is a constant proportion of that stimulus."[123] Because of his contributions to the further examination and dissemination of the principle, a modified version of Weber's law is known as Fechner's law today.[124] Fechner's first discussions with Wundt about the topic occurred after a young Georg Elias Müller had sharply criticized Fechner's account of Weber's law in *Elemente der Psychophysik*.[125] Müller was not a member of Wundt's Leipzig circle, but further discussions about the topic would primarily focus on the work done in Institute for Experimental Psychology, where Weber's law provided an important interpretive framework for investigations into almost all forms of perception.

Fechner did not only have a close relationship with Wundt but also discussed Weber's law with Wundt's collaborators.[126] They did not, however, use these short lines of communication with Fechner to have their work evaluated by him before sharing it with a wider audience. Instead, it was Fechner who used his good relationship with Wundt and his students to make sure that his public comments on what he perceived to be their misunderstanding of Weber's law would be appropriate. He shared his thoughts with the people he was criticizing for at least two reasons. On the one hand, he realised that a good personal relationship could be damaged by unexpected harsh criticism. By sharing this criticism in advance, Fechner not only prepared the recipients for the blow, but he also gave them the opportunity to make corrections before his comments would be published. If they decided not to correct anything, he had at least given them the opportunity to prepare a response.[127] On the other hand, Fechner used prepublication discussions to improve his own argumentation. He, for example, sent Wundt a draft paper in which he criticized his collaborators Volkmar Estel and Gustav Lorenz and asked Wundt to comment on his objections and counterarguments.[128] When he was preparing a critical paper about the work of another Wundt associate, Max Mehner, Fechner shared an early version of this paper with both Wundt and Mehner for this purpose, as well.[129]

Collaborative Work and Private Correspondence

Wundt's correspondence hardly shows any in-depth prepublication evaluation by outsiders of work carried out in his laboratory.[130] This is in stark contrast with the extensive discussions about the diphtheria serum between Behring and Heubner, which demonstrate that collaborative efforts can be evaluated in a meaningful way in private correspondence. The lack of such evaluation in Wundt's correspondence can be explained by a combination of circumstances.

One of these circumstances could also be observed in the earlier discussion of the diphtheria serum: both the serum and the findings of Wundt and his associates were the product of collaboration in a laboratory setting. In both cases, it was therefore highly unlikely that an individual commentator would be able to evaluate the findings through reiteration of the full intellectual process that had yielded them. None of Wundt's observations could easily be characterized as correct or incorrect in the same decisive way as this could be done for violations of the rules of grammar, rhyme, or metre in an edition of an Arabic manuscript. The fact that the observations at the Institute for Experimental Psychology were usually described as self-observation (*Selbstbeobachtung*), inner observation (*innere Beobachtung*), and inner experience (*innere Erfahrung*) made it even more unfeasible for outside reviewers to assess the raw data in a meaningful way.[131]

At the same time, there were no compelling reasons to submit Wundt's findings to an evaluation along the lines of the clinical trials of Behring's and Wernicke's serum. The replication of the data and production processes that had led to the serum had never been a major concern in these clinical tests. Instead, Heubner fully restricted himself to assessing the efficacy of the serum. The findings from Wundt's laboratory, however, were descriptions and analyses of physiological processes that could not be expected to have any kind of easily testable applicability. The template of the clinical trial could therefore not be used as the basis of any meaningful evaluation of these findings.

In one way, the data from Wundt's laboratory were easier to evaluate than Koch's and Behring's discoveries. Koch's *Tuberkulin* and Behring's diphtheria serum were both presented as revolutionary new findings, the manufacturing process and efficacy of which could not be traced back to that of earlier curative agents.[132] Most of the work in Wundt's laboratory, however, built on earlier findings of physiologists and psychophysicists. A reviewer in the *Literarisches Centralblatt*, for example, did not praise Wundt's *Grundzüge der physiologische Psychology* for its groundbreaking new findings, but rather for how it corresponded "exactly to the need created by recent developments in physiology and psychology."[133] One obvious way to evaluate the findings from Wundt's laboratory was therefore to assess if they corresponded with other researchers' findings. This was exactly what Fechner did. In his correspondence with Wundt and his associates, he not only compared their findings to what Weber's law predicted, he also reflected on the likely distribution of errors of measurement in their data sets and compared it to what he called "Gauss's law."[134] When the distribution of errors seemed to deviate from what, today, is known as the "Gaussian" or "normal" distribution, Fechner asked for further clarification.[135]

The letters exchanged between Wundt and Fechner demonstrate that a critically independent assessment of the findings from the former's laboratory in private correspondence was possible. It is curious, however, that Fechner was the only person to discuss these topics so extensively and candidly with Wundt. A closer look at the way in which the complex relation between loyal collegiality and critical independence were experienced by Wundt and his associates makes this remarkable observation more understandable.

One reason for the apparent insularity of the Institute for Experimental Psychology is rooted in the social character of laboratory collaboration. At this workplace, researchers can expect a fair amount of criticism and evaluation from their day-to-day colleagues, which can encourage the idea that outside evaluation would be superfluous. When he discussed Weber's law with Fechner, Wundt frequently underlined the trustworthiness of the findings published by his collaborators by emphasizing the fact that other members of his institute had repeated the experiments on which they were based: Julius Merkel and Gustav Lorenz, for example, had replicated experiments conducted by Ernst Tischer, while Volkmar Estel built on experiments done by Julius Kollert.[136] Some years later, Fechner mentioned another scholar of Wundt's institute—he did not remember his name—who had conducted experiments to verify Estel's work.[137]

This laboratory collaboration not only contributed to lessening the perceived need for further outside evaluation, but it also fostered an environment in which mutual loyalty among the collaborators was highly valued. Laboratory workers tended to look at their experimental findings as shared accomplishments. Therefore, Wundt often fiercely defended his collaborators against Fechner's criticism. In most of his replies, his polite rhetorical strategy was not to claim that Fechner was wrong, but to clear up apparent misunderstandings.[138] Meanwhile, Fechner continuously emphasized that he kept harbouring doubts and could therefore not agree with Wundt's objections.[139] As in their earlier discussion of Slade's spiritism, both men generally had to settle for an agreement to disagree.

Although the above considerations of loyalty to one's everyday colleagues was shared by researchers in all types of laboratories, the character of the experiments at Wundt's institute provided an additional reason to be suspicious of outsiders. The experiments generally required people to take on three different roles, namely those of experimenter, observer, and subject. The people in all roles were expected to have a clear understanding of the experiment.[140] In this setting, the role of subject was considered to require "more psychological sophistication" than that of the experimenter or observer.[141] This requirement arose from the type of questions

Wundt and his students would ask, as these were usually aimed at establishing the minimal perceivable difference between two impulses or the minimal reaction time in response to a stimulus. In order to perceive this minimal difference or to keep reaction times as short as possible, subjects had to be well-trained.[142] Most members of the Institute would alternate between the role of experimenter, observer, and subject: an efficiently run laboratory for experimental psychology could not afford to use well-trained people in one role only. However, this experimental set-up did not foster trust in the judgement of outsiders. Even if they would understand what it took to be an experimenter or observer, they could hardly be expected to have access to subjects who were trained to be sources of meaningful data.

In addition to the characteristics of laboratory cooperation, there were personal qualities that prevented Wundt from benefiting from the supportive evaluation of his not-yet-published work. The case of Nöldeke and De Goeje illustrates that the willingness of peers to invest in each other's not-yet-published works can be encouraged by a close personal relationship. Wundt, however, had not developed intimate and lasting friendships with the future medical doctors and physiologists with whom he had studied in Tübingen and Heidelberg. His correspondence lacks letters from former fellow students and the section about his student days in his autobiography does not mention lasting friendships with them.[143]

This was not because he was unable to engage in personal relationships. In the same book, he gratefully looks back at the relationship with his high school friends Heinrich Holtzmann and Adolf Hausrath.[144] Over the decades, these three men would share memories and witticisms. Holtzmann, for example, remembered Wundt of their time at an "obscure ale-house in Karlsruhe" which he called "the true university and everything your heart desires."[145] Hausrath joked about Wundt's style of thought and his election to the *Académie Française* in 1904: "I would like to warmly congratulate you with your acceptance among the French immortals. May you succeed in completely reconciliating this great nation with us [Germans]. Without a doubt you will construct an apparatus that will reduce the [...] French and German yearning for revenge to a pure mathematical formula."[146] Like Wundt, Holzmann and Hausrath had successful academic careers. However, because they both became theologians, they were unable to contribute to their friend's work in a meaningful way.

Wundt and his collaborators hardly received any supportive feedback from the members of other philosophy departments, either.

Late-nineteenth-century German philosophy departments consisted primarily of Kantians and Hegelians, who looked at Wundt's efforts with suspicion. Some of his colleagues thought that the attempt to use laboratory experiments to bridge the gap between the natural sciences and philosophy could only lead to materialist, unchristian, and un-German results. Richard Avenarius, the editor of a journal to which Wundt regularly contributed, was considered unfit for a Prussian professorship because he was described to Althoff as the representative of "a very extreme school of thought."[147] Others saw Wundt's experimental contributions as so new and underdeveloped that they could not yet offer a reliable template for further research. Even those who praised Wundt assumed that others would not be able to succeed in a similar fashion.[148] In addition, Wundt was bad at maintaining cordial relationships with the few people who would have been able to provide meaningful evaluations of his work. The Halle university librarian Otto Hartwig reported to Althoff that Wundt had fallen out with all "the greatest" scholars in Berlin, especially with his former employer Helmholtz.'[149]

All in all, the lack of evaluation of works from the Institute for Experimental Psychology can be explained by several factors. The social dynamics that Wundt's laboratory shared with other laboratories are one part of the explanation. The special character of the research methods at his laboratory is another part. In addition, Wundt's personal relationships with other scholars and his relative intellectual isolation in late-nineteenth-century German philosophy provide another clue to the relative lack of evaluative comments in his correspondence. Wundt had close relationships with scholars who were unable to give useful feedback on his work and maintained uneasy relationships with people whose insights could have been useful. Of the scholars outside of Wundt's own institute, only Fechner regularly discussed issues of shared academic interest with him, and these discussions tended to lead to little more than an ever-recurring agreement to disagree.

Balancing Loyalty and Independence in Private Correspondence

The above case studies share one very characteristic feature—they all focus on the evaluation of scholarly work by other scholars before sharing it with a broader audience. In their correspondences, all the protagonists had to deal with the demands and expectations evoked by their recognition of the importance of loyal collegiality and critical independence. The case studies have illustrated that there was not one way to manage the potential coexistence of and tensions between these virtues that was shared by individuals with different disposition working in different disciplines. Their choices were not only shaped by their appreciation of collegial loyalty and critical independence, the characteristics of the practice of mutual evaluation through private correspondence also shaped their balancing act.

One of these characteristics was its non-mandatory character. No scholar had to feel obliged to evaluate someone else's work. Although it was not uncommon for scholars to feel obliged to review the work of friends, acquaintances, admired colleagues, or former students, there was no obligation to invest a lot of time and effort into the support of relative strangers. This meant that respectable, yet isolated scholars did not benefit from the advantages of peer evaluation. Wilhelm Ahlwardt, orientalist in Greifswald, was one of them. Nöldeke and De Goeje often criticized his catalogues and text editions. Nöldeke blamed the decreasing quality of his work on his isolation: "Ahlwardt is a curious, *lonely* fellow. If he would have had a closer relationship with his peers, he would have done things differently. [...] Forty years ago, Ahlwardt certainly was the best expert on Arabic poetry, but he has hardly learned anything new since that time and there is probably also much that he has forgotten. (Oh, how erudite would we be, if we could remember everything we ever knew!!!)."[150]

A second characteristic is the fact that meaningful evaluation was intellectually demanding and time-consuming. For many years, Nöldeke thoroughly checked not only thousands of pages of Arabic texts edited by De Goeje, but he also did this favour to others. Meanwhile, De Goeje did the same for Nöldeke and other peers. Heubner's investment in the diphtheria serum was very demanding, as well. The first letters mentioning his tests for Behring were sent in 1892 and four years later he was still involved with them.[151] And he was not the only one engaged in such tests; the Berlin doctors Eduard Henoch and Ernst von Bergmann and the Munich physician Max Joseph Oertel spent considerable time on the testing of Behring's serum too.[152] Finally, Fechner's extensive correspondence about the work carried out in Wundt's laboratory was very time-consuming as well. His letters often counted dozens of pages and the longest even counted as much as 121 pages![153]

Another characteristic of private evaluation was its perceived usefulness. Sometimes, it could be useful to the evaluator himself. This was the case with Otto Heubner, who was motivated by his experiences of being unable to cure children from diphtheria. If he could help to advance the development of the new serum, this could benefit both the sick children and himself. After all, he would have done an outstanding job as a paediatrician if he proved to be able to contribute to the eradication of this deadly disease. The people who were most likely to profit from private evaluative efforts, however, were the scholars whose work was evaluated. Heubner contributed both the necessary knowledge and the indispensable resources that Behring lacked, which consisted of a long experience in observing and diagnosing diphtheria in people, the competence to manage large-scale clinical research and a statistically significant number of diphtheria patients.

Another notable feature of private evaluation is that different products of scholarship called for a different form of evaluation. Texts are ideally suited for exhaustive mutual evaluation. Therefore, Nöldeke and De Goeje were able to thoroughly review each other's work. Research results, however, do not always consist of texts. The most important outcome of Behring's and Wernicke's endeavours was a serum. In the case of Wundt and his collaborators, their primary results consisted of a series of measurements. These results did not lend themselves to peer evaluation as easily as texts because they were the result of collaboration. The complexity of such a collaborative effort could not be replicated as easily as the thoughts and considerations of an individual philologist. The findings from Wundt's laboratory in particular suffered from replication problems because they were the result of the intensive training of experimenters, observers, and subjects. This contributed to a laboratory culture in which mutual evaluation largely took place within the walls of the institute, while criticism from the outside was kept at a distance.

A final defining feature of the practice of private evaluation is its informal character. Among close acquaintances, it could therefore be difficult to refuse to evaluate each other's work. De Goeje made long lists of suggestions for numerous people, with whom he was not particularly close, such as Ferdinand Wüstenfeld, Louis Cheikho, and Carl Brockelmann.[154] The fact that De Goeje did not even refuse his time and energy to these rather distant acquaintances suggests that it would have been unthinkable to refuse to support Nöldeke. In a similar fashion, the extensive and detailed correspondence between Fechner and Wundt cannot be explained by simply pointing at the fact that they shared a professional interest. Though only one letter from Fechner to Wundt survived from before his call to Leipzig, by far the largest part of their correspondence took place during Wundt's Leipzig years. The fact that the two men had become friends in real life added to their willingness to extensively discuss shared interests.

Some of the characteristics of private evaluation encouraged collegial loyalty. Its non-binding character vis-à-vis relative strangers and the hard work it generally involved suggested a benevolent goodwill when a scholar accepted the responsibility of assessing his peer's work. The strong pressure to support friends and close colleagues also suggests that private evaluation is a setting in which loyalty is a primary motivation. At the same time, however, private assessments of research could only be useful if the evaluation would be made by a critical and independent peer. The way in which a meaningful critical stance could be asserted depended on the subject matter of the research and the extent to which it was the result of a collaborative effort.

Often, the shared acknowledgement of both the informal character of the evaluation and the importance of a critically independent assessment went hand in hand. In the case of Nöldeke and De Goeje, the friendship between them allowed the exercise of independent judgement as an expression of loyalty. In the case of Behring and Heubner, a shared goal allowed the exercise of independent judgement as a demonstration of loyal collegiality. But such an alignment of both values was not attainable for everyone. Wundt's few knowledgeable friends were not as close to him as his everyday collaborators and in the absence of any overarching goals shared with people outside of his inner circle, collegial loyalty and critical independence did not go together in any obvious way. He remained loyal to his everyday collaborators and kept a critical distance from others.

Notes

1. Steven Harris, for example, mentions that the golden age of the Republic of letters was in the eighteenth century: Steven J. Harris, "Networks of Travel, Correspondence and Exchange," in *The Cambridge History of Science*, Volume 3, ed. Katharine Park and Lorraine Daston (Cambridge: Cambridge University Press, 2006), 349. Ian McNeely's and Lisa Wolverton's chapter on the Republic of Letters explicitly deals with the period between 1500 and 1800: Ian F. McNeely and Lisa Wolverton, *Reinventing Knowledge: From Alexandria to the Internet* (New York, NY: Norton, 2008), Chapter 4. The importance of scholarly correspondence in the German lands in the seventeenth and eighteenth century is high-

lighted in: Klaus-Dieter Herbst and Stefan Kratochwil (eds.), *Kommunikation in der Frühen Neuzeit* (Frankfurt am Main: Peter Lang, 2009) and Ulrich Johannes Schneider, (ed.), *Kulturen des Wissens im 18. Jahrhundert* (Berlin: Walter de Gruyter, 2008).
2. Peter Burke, "The Republic of Letters as a communication system: An essay in periodization," *Media History*, 18, no. 3–4 (2012): 397–398.
3. Ibid.
4. Franz Mauelshagen, "Networks of Trust: Scholarly Correspondence and Scientific Exchange in Early Modern Europe," *The Medieval History Journal*, 6, no. 1 (2003): 2.
5. Richard van Dülmen, *Die Entdeckung des Individuums, 1500–1800* (Frankfurt am Main: Fischer, 1997), 106 and Joachim Kirchner, *Das deutsche Zeitschriftwesen: seine Geschichte und seine Probleme*, Teil 1, 2. neu bearbeitete und erweiterte Auflage (Wiesbaden: Harrasowitz, 1958), 14.
6. Peter Burke, "The Republic of Letters," 398.
7. Wolf Feuerhahn and Pascale Rabault-Feuerhahn, "Présentation: la science à l'échelle internationale," *Revue germanique internationale* 12 (2010): 5.
8. Louise Miskell, *Meeting Places: Scientific Congresses and Urban Identity in Victorian Britain* (Farnham: Ashgate, 2013), 2.
9. Holger Preissler, "Die Anfänge der Deutschen Morgenländischen Gesellschaft," *Zeitschrift der Deutschen Morgenländischen Gesellschaft* 145, no. 2 (1995): 241–327.
10. This theme runs throughout Mauelshagen's "Networks of Trust" (see note 3).
11. Erika Krauße, "Vorbemerkung: Der Brief als wissenschaftshistorische Quelle," in *Der Brief als wissenschaftshistorische Quelle*, ed. Erika Krauße (Berlin: Verlag für Wissenschaft und Bildung, 2005), 6.
12. Ton van Kalmthout, "Scholarly Correspondence and the History of Philology: A Case Study of the Nineteenth-Century Netherlands," *Amsterdamer Beiträge zur älteren Germanistik*, 78 (2018), 155.
13. William M. Calder III and Bernhard Huss, "Introduction," in *"Sed serviendum officio …" The Correspondence between Ulrich von Wilamowitz-Moellendorff and Eduard Norden (1892–1931)*, ed. William M. Calder III and Bernhard Huss (Hildesheim: Weidmann, 1997), xii.
14. Armin Hermann, "Die Funktion von Briefen in der Entwicklung der Physik," *Berichte zur Wissenschaftsgeschichte* 3, no. 1/2 (1980): 63.
15. Gerhard Wagner and Gilbert Weiss, "Editors' Introduction," in *A Friendship That Lasted a Lifetime: The Correspondence Between Alfred Schütz and Eric Voegelin*, translated by William Petropulos, ed. Gerhard

Wagner and Gilbert Weiss (Columbia, MO: University of Missouri Press, 2011), 3.

16. The following biographical sketch draws on the introduction in Bernhard Maier, *Gründerzeit der Orientalistik: Theodor Nöldekes Leben und Werk im Spiegel seiner Briefe* (Würzburg: Ergon Verlag, 2013).
17. Christiaan Engberts, "Gossiping about the Buddha of Göttingen: Heinrich Ewald as an Unscholarly Persona," *History of Humanities* 1, no. 2 (2016): 371–385.
18. Theodorus Nöldeke, *De origine et compositione Surarum qoranicarum ipsiusque Qorani* (Göttingen: Verlag der Dieterichschen Buchhandlung, 1856).
19. The following biographical sketch draws on the portrait in: Arnoud Vrolijk and Richard van Leeuwen, *Arabic Studies in the Netherlands: A Short History in Portraits, 1580–1950* (Leiden: Brill, 2014).
20. Christiaan Snouck Hurgronje, "Michaël Jan de Goeje," in *Jaarboek der Koninklijke Akademie van Wetenschappen* (Amsterdam, 1909), 117.
21. Universiteitsbibliotheek Leiden (hereafter UBL): BPL 2389, Michael Jan de Goeje to Theodor Nöldeke, no date, written as a reply to a letter by Nöldeke of 14 October 1907.
22. UBL: Or: 8952 A: 770, Theodor Nöldeke to Christiaan Snouck Hurgronje, 3 March 1923.
23. Suzanne L. Marchand, *German Orientalism in the Age of Empire: Religion, Race, and Scholarship* (Cambridge: Cambridge University Press, 2009), 175; Theodor Nöldeke, *Geschichte des Qorāns* (Göttingen: Verlag der Dieterichschen Buchhandlung, 1860).
24. For Berlin see: UBL: BPL 2389, Theodor Nöldeke to Michael Jan de Goeje, 28 February 1875. For Berlin see UBL: BPL 2389, Theodor Nöldeke to Michael Jan de Goeje, 20 June 1888. For Vienna see UBL: BPL 2389, Theodor Nöldeke to Michael Jan de Goeje, 15 December 1879.
25. UBL: BPL 2389, Michael Jan de Goeje to Theodor Nöldeke, 13 August 1864.
26. UBL: BPL 2389, Theodor Nöldeke to Michael Jan de Goeje, 14 October 1907.
27. UBL: BPL 2389, Michael Jan de Goeje to Theodor Nöldeke, 12 June 1866.
28. UBL: BPL 2389, Michael Jan de Goeje to Theodor Nöldeke, 2 February 1904.
29. UBL: BPL 2389, Theodor Nöldeke to Michael Jan de Goeje, 20 June 1888. Nöldeke's emphasis.
30. UBL; BPL 2389, Michael Jan de Goeje to Theodor Nöldeke, 14 November 1896.

31. UBL: BPL 2389, Michael Jan de Goeje to Theodor Nöldeke, 9 December 1863, Theodor Nöldeke to Michael Jan de Goeje, 4 February 1864 and 1 Mai 1864.
32. UBL: BPL 2389, Michael Jan de Goeje to Theodor Nöldeke, 13 December 1865; Theodor Nöldeke, 'Historia khalifatus Omari II, Jazídi II et Hischámi,' *Göttingische gelehrte Anzeigen*, 44, 1865, 1747–1753.
33. Ghada Osman, "Oral vs. Written transmission: The case of Tabarī and Ibn Sa'd," *Arabica* 48, no. 1 (2001): 66.
34. Franz-Christoph Muth, *Die Annalen von aṭ-Ṭabarī im Spiegel der europäischen Bearbeitungen* (Frankfurt am Main: Peter Lang, 1983), 1–3.
35. Joannes Godefredus Ludovicus Kosegarten (ed.), *Annales regum atque legatorum dei ex codice manuscript berolinensi* (Greifswald: Ernst Mauritius, 1831–1853).
36. Snouck Hurgronje, "Michaël Jan de Goeje," 129.
37. Michael Jan de Goeje, "Literarische Notiz," *Zeitschrift der Deutschen Morgenländischen Gesellschaft* 16 (1862), 759–762.
38. UBL: BPL 2389, Michael Jan de Goeje to Theodor Nöldeke, 25 August 1865.
39. UBL: Or. 5585e, Albert Socin to Michael Jan de Goeje, 22 December 1872.
40. For example: UBL: BPL 2389, Michael Jan de Goeje to Theodor Nöldeke, 17 March 1873; UBL: Or. 5585e, Arist Aristovich Kunik to Michael Jan de Goeje, 27 February 1873, and Eduard Sachau to Michael Jan de Goeje, 10 May 1873.
41. UBL: Or. 5585e Otto Loth to Michael Jan de Goeje, 22 February 1874; UBL: BPL 2389, Theodor Nöldeke to Michael Jan de Goeje, 7 April 1874.
42. UBL: Or. 5585e, Eduard Sachau to Michael Jan de Goeje, 23 October 1873, Andreas David Mordtmann to Michael Jan de Goeje, 6 March 1874.
43. On the establishment of the consortium, see: Christiaan Engberts, "Een taalkundige als internationaal ondernemer: M.J. de Goeje en de Annalen van al-Tabari (1872–1901)," *De Moderne Tijd* 4, no. 1/2 (2020): 36–54.
44. Theodor Nöldeke, *Geschichte der Perser und Araber zur Zeit der Sasaniden: aus der Arabischen Chronik des Tabari übersetzt und mit ausführlichen Erläuterungen und Ergänzungen versehn* (Leiden, Brill, 1879).
45. UBL: Or. 5585f, Theodor Nöldeke to Michael Jan de Goeje, 21 June 1876.
46. UBL: Or. 5585f, Allgemeine Bestimmungen für die Herausgabe des Tabarî; UBL: Or. 5585f, Theodor Nöldeke to Michael Jan de Goeje, 21 June 1876.
47. On Ignazio Guidi: UBL: 5585f, undated notes in Theodor Nöldeke's handwriting about the second series of al-Ṭabarī's *Annals*, 978–1275;

On Pieter de Jong: UBL 5585f, Theodor Nöldeke to Michael Jan de Goeje, 26 September 1876.
48. Theodor Nöldeke to Eduard Meyer, 27 July 1907, in: Audring, G., Der Briefwechsel zwischen Theodor Nöldeke und Eduard Meyer.
49. Not all of his lists of comments are dated. Some of the ones that have been dated are: UBL: Or. 5585f, Theodor Nöldeke to Michael Jan de Goeje, 1 April 1885 and 4 October 1896.
50. UBL: Or. 5585f, Theodor Nöldeke to Michael Jan de Goeje, 20 February 1876. Nöldeke's emphasis.
51. For comments on the *sukūn*, see: UBL: Or. 5585f, undated list of comments on series III, pages 1923–2278; for comments on the *tanwīn*, see: UBL: Or. 5585f, undated list of comments on series II, pages 978–1275.
52. Boaz Shoshan, *Poetics of Islamic Historiography: Deconstructing Ṭabarī's History* (Leiden: Brill, 2004), xxxi.
53. Ghada Osman, "Oral vs. Written Transmission," 66–67.
54. A.F.L. Beeston and Lawrence I. Conrad, "On Some Umayyad Poetry in the History of al-Ṭabarī," *Journal of the Royal Asiatic Society*, Series 3, 3, no. 2 (1993): 191.
55. UBL: Or. 5585f, Theodor Nöldeke to Michael Jan de Goeje, 13 October 1875.
56. UBL: Or. 5585f, Theodor Nöldeke to Michael Jan de Goeje, 9 February 1876.
57. UBL: Or. 5585f, Theodor Nöldeke to Michael Jan de Goeje, 17 March 1891; UBL: Or. 5585f, undated notes in Theodor Nöldeke's handwriting about the second series of al-Ṭabarī's *Annals*, 978–1275.
58. UBL: Or. 5585f, Theodor Nöldeke to Michael Jan de Goeje, 1 April 1885.
59. Boaz Shoshan, *Poetics of Islamic Historiography*, xxix.
60. A.F.L. Beeston and Lawrence I. Conrad, "On Some Umayyad Poetry," 191.
61. Sir Charles Lyall in his *Ancient Arabian Poetry* as quoted in Reynold A. Nicholson, *A Literary History of the Arabs*, Second Impression (London: T. Fischer Unwin, 1914), 75.
62. UBL: Or. 5585f, Theodor Nöldeke to Michael Jan de Goeje, 26 September 1876.
63. UBL: Or. 5585f, Theodor Nöldeke to Michael Jan de Goeje, 26 September 1876 and 4 October 1896.
64. UBL: Or. 5585f, undated notes in Theodor Nöldeke's handwriting about the second series of al-Ṭabarī's *Annals*, 978–1275 and undated notes in Theodor Nöldeke's handwriting about the second series of al-Ṭabarī's *Annals*, 606–933.
65. UBL: Or. 5585f, Theodor Nöldeke to Michael Jan de Goeje, 1 April 1885.

66. UBL: Or. 5585f, Theodor Nöldeke to Michael Jan de Goeje, 26 September 1876.
67. A.F.L. Beeston and Lawrence I. Conrad, 'On Some Umayyad Poetry,' 192.
68. UBL: 5585f, Theodor Nöldeke to Michael Jan de Goeje, 26 September 1876 and 4 October 1896.
69. UBL: BPL 2389, Michael Jan de Goeje to Theodor Nöldeke, 25 August 1865.
70. UBL: BPL 2389, Michael Jan de Goeje to Theodor Nöldeke, 11 February 1903.
71. UBL: Or. 8952 A: 763, Theodor Nöldeke to Christiaan Snouck Hurgronje, 13 January 1918.
72. Magyar Tudományos Akadémia Könyvtár Információs Központ (Library and Information Centre of the Hungarian Academy of Sciences, hereafter MTAK), GIL/32/01/156, Theodor Nöldeke to Ignaz Goldziher, 4 April 1904.
73. UBL: Or. 5585f, Theodor Nöldeke to Michael Jan de Goeje, 26 September 1876 and 4 October 1896.
74. P. Kleist and C. Zerobin Kleist, "Eine kurze Geschichte der klinischen Studie," *Schweizerische Ärztezeitung* 86, no. 44 (2005): 2477; Abraham M. Lilienfeld, "Ceteris paribus: the evolution of the clinical trial," *Bulletin of the History of Medicine* 56, no. 1 (1982): 4–5; Antje Timmer, "Kontrollierte klinische Studien vor Archie Cochrane," *Zeitschrift für Evidenz, Fortbildung und Qualität im Gesundheitswesen* 102 (2008): 474–475.
75. Robert Koch, "Weitere Mittheilungen über ein Heilmittel gegen Tuberkulose," *Deutsche Medizinische Wochenschrift* 16 (1890): 1029–1032.
76. Barbara Elkeles, "Der »Tuberkulinrausch« von 1890," *Deutsche Medizinische Wochenschrift*, 115, no. 45 (1990): 1729.
77. Christoph Gradmann, "A harmony of illusions: clinical and experimental testing of Robert Koch's tuberculin 1890–1900," *Studies in History and Philosophy of Biological and Biomedical Sciences* 35 (2004): 474.
78. Christoph Gradmann, "Robert Koch and the Pressures of Scientific Research: Tuberculosis and Tuberculin," *Medical History* 45 (2001): 24–25.
79. For example, see the letter of 27 October 1890 von Gustav von Goßler to the Kaiser and the opinion of chancellor Leo von Caprivi of 25 December1890, both in: GStA PK, VI. HA, Nl Althoff, No.313: Das Kochsche Tuberculin.
80. This short overview of Behring's life draws from: Derek S. Linton, *Emil von Behring: Infectious Disease, Immunology, Serum Therapy* Philadelphia,

PA: American Philosophical Society, 2005) and H. Zeiss and R. Bieling, *Behring: Gestalt und Werk* (Berlin-Grunewald: Bruno Schultz, 1940).

81. More about Wernicke can be found in Erika Schulte, *Der Anteil Erich Wernickes an der Entwicklung des Diphtherieantitoxins* (Berlin: Mensch & Buch, 2001).
82. Erich Wernicke to Thorwald Madsen, [no date], Behring-Nachlass digital (hereafter BNd): EvB/F5/1.
83. Ibid.
84. Emil Behring to Otto Heubner, 8 May 1893, BNd: EvB/B1/54. All emphases by Behring.
85. Erich Wernicke to Thorwald Madsen, [no date], BNd, EvB/F5/1.
86. H. Zeiss and R. Bieling, *Behring: Gestalt und Werk*, 103.
87. These first tests are mentioned in: Erich Wernicke to Thorwald Madsen, [no date], BNd: EvB/F5/1. Zeiss and Bieling say they are “a legend”: H. Zeiss and R. Bieling, *Behring: Gestalt und Werk*, 85. Based on the letter to Madsen and some notes Wernicke added to a picture of Behring and himself, Oedingen and Staerk support December 1890 as the moment of the first test of the serum on people: Christina Oedingen and Joseph W. Staerk, “First Cure for Diphtheria by Antitoxin as Early as 1891,” *Annals of Science* 54 (1997): 607–610. Linton is inconclusive: Linton, *Emil von Behring*, 112–116.
88. Heubner’s biographical details have been drawn from Heinz Goerke, *Berliner Ärzte: Virchow, Graefe, Koch, Leyden, Bergmann, Bier, Heubner, Moll, Stoeckel*, zweite Auflage (Berlin: Berlin Verlag, 1984), 192–198.
89. Emil Behring to Otto Heubner, 20 July 1892, BNd: EvB/B1/57.
90. August Laubenheimer to Aufsichtsrat der Farbwerken Vormals Meister, Lucius & Brünning, 20 April 1894, BNd: EvB/B196/13.
91. Emil Behring to Otto Heubner, 11 November 1892, 23 December 1892, and 12 April 1893: BNn: EvB/B1/41, EvB/B1/46, and EvB/B1/52.
92. August Laubenheimer to Aufsichtsrat der Farbwerken Vormals Meister, Lucius & Brünning, 20 April 1894, BNd: EvB/B196/13.
93. Emil Behring to Otto Heubner, 9 April 1893, Sammlung Darmstaedter in der Staatsbibliothek Berlin (hereafter Slg. Darmstaedter): 3a 1890 (5) and Emil Behring to Otto Heubner, 9 May 1893, BNd: EvB/B1/56.
94. Emil Behring to Otto Heubner, 4 May 1893, BNd: EvB/B1/53.
95. Emil Behring to Otto Heubner, 8 May 1893, BNd: EvB/B1/54.
96. Emil Behring to Otto Heubner, 5 December 1892, 23 December 1892, and 4 4 May 1893, BNd: EvB/B1/42, EvB/B1/46, and EvB/B1/53.
97. Emil Behring to Otto Heubner, 5 December 1892, BNd: EvB/B1/42.
98. Emil Behring to Otto Heubner, 12 April 1893, BNd: EvB/B1/52.
99. Emil Behring to Otto Heubner, 8 May 1893, BNd: EvB/B1/55.

100. Report from August Laubenheimer, copied in: Carl Ludwig Lautenschläger to das Behring-Archiv, Marburg, Alexander von Engelhardt, 14 February 1941, BNd, EvB/B 196/7.
101. Axel C. Hüntelmann, "Diphtheria serum and serotherapy. Development, Production and regulation in *fin de siècle* Germany," *Dynamis*, 27 (2007): 113–114.
102. Timothy Lenoir, "A Magic Bullet: Research for Profit and the Growth of Knowledge in Germany Around 1900," *Minerva* 26, no. 1 (1988): 71–74.
103. Erich Wernicke to Bernhard Möllers, 29 August 1924, BNd, EvB/F5.
104. August Laubenheimer to Aufsichtsrat der Farbwerken Vormals Meister, Lucius & Brünning, 20 April 1894, BNd: EvB/B196/13. For more on Ehrlich's contribution to the diphtheria serum, see Ernst Bäumler, *Paul Ehrlich: Forscher für das Leben*, 3., durchgesehene Auflage (Frankfurt am Main: Edition Wötzel, 1997), 92–93 and Axel C. Hüntelmann, *Paul Ehrlich: Leben, Forschung, Ökonomien, Netzwerke* (Göttingen: Wallstein Verlag, 2011), 91–108.
105. Beratung betreffend das Diphterieserum, [o. Datum; sicher aber am 3.11.1894), BNd, EvB/B196/5.
106. Derek S. Linton, *Emil von Behring*, 179.
107. For example: Emil Behring, *Die Blutserumtherapie I: Die praktischen Ziele der Blutserumtherapie und die Immunisierungsmethoden zum Zweck der Gewinnung von Heilserum* (Leipzig: Georg Thieme, 1892) and Emil Behring and Erich Wernicke, "Ueber Immunisirung und Heilung von Versuchsthieren bei der Diphtherie," *Zeitschrift für Hygiene* 12 (1892): 45–57.
108. Ernst Bäumler, *Paul Ehrlich*, 93 and Axel C. Hüntelmann, "The Dynamics of *Wertbestimming*,' *Science in Context* 21, no. 2 (2008): 229–252.
109. One of these unpaid assistants was Wernicke's fiancée, Meta Füth, who, while living in Friedeberg, not only fed some sheep bought by Wernicke, but also injected them with diphtheria. Much of the serum that Heubner received was produced under her supervision. Erich Wernicke to Thorwald Madsen, [no date], BNd: EvB/F5/1.
110. See page 16.
111. The overview of Wundt's early years primarily draws from: Wolfgang G. Bringmann, Norma J. Bringmann and William D.G. Balance, "Wilhelm Maximilian Wundt 1832–1874: The Formative Years," in *Wundt Studies: A Centennial Collection*, ed. Wolfgang G. Bringmann and Ryan D. Tweney (Toronto: C.J. Hogrefe, 1980), 13–32 and Solomon Diamond, "Wundt before Leipzig," in *Wilhelm Wundt in History: The Making of a Scientific Psychology*, ed. Robert W. Rieber and David

K. Robinson (New York, NY: Springer Science + Business Media, 2001), 1–68.

112. Georg Lamberti, *Wilhelm Maximilian Wundt (1832–1920): Leben, Werk und Persönlichkeit in Bildern und Texten* (Bonn: Deutscher Psychologen Verlag, 1995), 23–39.
113. Wolfgang G. Bringmann et al., "Wilhelm Maximilian Wundt," 27.
114. Universitätsarchiv Leipzig (hereafter UAL), Nl. Wilhelm Wundt, Wilhelm Wundt to Sophie Mau, 15 June 1872.
115. Wilhelm Wundt, *Grundzüge der Physiologischen Psychologie* (Leipzig: Engelmann, 1874).
116. Wilhelm Wundt, "Das Institut für experimentelle Psychologie," in *Festschrift zur Feier des 500 jährigen Bestehens der Universität Leipzig*, 4. Band, I. Teil, ed. Rektor und Senat der Universität Leipzig (Leipzig: S. Hirzel, 1909), 118.
117. The argument for the establishment of Wundt's Leipzig laboratory as central to the establishment of psychology as an autonomous science is made in: Joseph Ben-David and Randall Collins, "Social factors in the origins of a new science: the case of psychology," *American Sociological Review* 31, no. 4 (1966): 451–465; Wolfgang G. Bringmann and Gustav A. Ungerer, "The Foundation of the Institute for Experimental Psychology at Leipzig University," *Psychological Research* 42 (Wundt Centennial Issue) (1980): 5–18. A critical counterargument has been made in Alexandre Métraux, "Wilhelm Wundt und die Institutionalisierung der Psychologie: Ein Beitrag zu einem kontroversen Kapitel der Psychologiegeschichte," *Psychologische Rundschau*, Band XXXI (1980): 84–98.
118. Klaus B. Staubermann, "Tying the knot; skill, judgement and authority in the 1870s Leipzig spiritistic experiments," *The British Journal for the History of Science* 34, no. 1 (2001): 73–74.
119. Ibid., 75.
120. Wilhelm Wundt, *Der Spiritismus: eine sogenannte wissenschaftliche Frage: offener Brief an Herrn Prof. Dr. Hermann Ulrici in Halle* (Leipzig: Wilhelm Engelmann, 1879).
121. UAL, Nl. Wilhelm Wundt, Gustav Theodor Fechner to Wilhelm Wundt, 18 June 1879.
122. UAL, Nl. Wilhelm Wundt, Gustav Theodor Fechner to Wilhelm Wundt, 25 June 1879.
123. Arthur L. Blumenthal, "Shaping a Tradition: Experimentalism Begins," in *Points of View in the Modern History of Psychology*, ed. Claude E. Buxton (Orlando, FL: Academic Press, 1985), 55.
124. The search result for 'Weber's law' at the online version of the Oxford Dictionary of Biomedicine simply states: 'see Fechner's law.' John Lackie,

Oxford Dictionary of Biomedicine, 2010, DOI: 10.1093/acref/9780199549351.001.0001. For a more detailed account of Fechner's contributions to Weber's law, see Michael Heidelberger, *Nature from within: Gustav Theodor Fechner and his psychophysical worldview*, translated by Cynthia Klohr (Pittsburgh, PA: University of Pittsburgh Press, 2004), 200–207.

125. Gustav Theodor Fechner, *Elemente der Psychophysik* (Leipzig: Breitkopf und Härtel, 1860). Müller most famously published his criticisms in Georg Elias Müller, *Zur Grundlegung der Psychophysik: Kritische Beiträge* (Berlin: Griebel, 1878). Fechner's discussion of Elias's criticism with Wundt can be found in UAL, Nl. Wilhelm Wundt, Gustav Theodor Fechner to Wilhelm Wundt, 27 October 1880 and 28 October 1880.
126. Fechner, for example, exchanged views with Gustav Lorenz and Max Mehner: UAL, Nl. Wilhelm Wundt, Gustav Theodor Fechner to Wilhelm Wundt, 12 July 1885 and 13 April 1886.
127. UAL, Nl. Wilhelm Wundt, Gustav Theodor Fechner to Wilhelm Wundt, 24 March 1885 and 12 July 1885.
128. UAL, Nl. Wilhelm Wundt, Gustav Theodor Fechner to Wilhelm Wundt, 7 April 1885.
129. UAL, Nl. Wilhelm Wundt, Gustav Theodor Fechner to Wilhelm Wundt, 13 April 1886.
130. Part of his correspondence with his Leipzig collaborators will be discussed in the next chapter: in these discussions Wundt figured in the role of editor of his own journal, the *Philosophische Studien*, who commented on the papers of his own collaborators.
131. David Kent Robinson, 'Wilhelm Wundt and the establishment of experimental psychology, 1875–1914: The context of a new field of scientific research,' (PhD diss., University of California, Berkeley, 1987). 100.
132. Even though it can be argued that the 'bacteriological scorched-earth strategy' that Koch assumed to explain the efficacy of Tuberkulin followed from pre-existing ideas about the relationship between disease and pathogen, it was a new and unique theory of the efficacy of a curative agent: Christoph Gradmann, "A harmony of illusions," 370–371. On the innovativeness of Behring's serum, see Derek S. Linton, *Emil von Behring*, 3–6.
133. Solomon Diamond, "Wundt before Leipzig," 59.
134. UAL, Nl. Wilhelm Wundt, Theodor Gustav Fechner to Wilhelm Wundt, 24 March 1885 and 11 April 1885.
135. On the messy history of this terminology, see Stephen M. Stigler and William H. Kruskal, "Normative Terminology," in Stephen M. Stigler, *Statistics on the Table: The History of Statistical Concepts and Methods* (Cambridge, MA: Harvard University Press, 1999), 403–430.

136. UAL, Nl. Wilhelm Wundt, Wilhelm Wundt to Gustav Theodor Fechner, 27 October 1882 and 10 November 1882.
137. UAL, Nl. Wilhelm Wundt, Gustav Theodor Fechner to Wilhelm Wundt, 14 April 1885.
138. UAL, Nl. Wilhelm Wundt, Wilhelm Wundt to Gustav Theodor Fechner, 19 October 1882 and 27 October 1882.
139. UAL, Nl. Wilhelm Wundt, Gustav Theodor Fechner to Wilhelm Wundt, 11 November 1882 and 11 April 1885.
140. David Kent Robinson, 'Wilhelm Wundt and the establishment of experimental psychology,' 101.
141. Kurt Danziger, *Constructing the subject: Historical origins of psychological research* (Cambridge: Cambridge University Press, 1990), 51.
142. David Kent Robinson, 'Wilhelm Wundt and the establishment of experimental psychology,' 101.
143. Wilhelm Wundt, *Erlebtes und Erkanntes* (Stuttgart: Alfred Kröner, 1920), 60–102.
144. Ibid., 55–56.
145. UAL, Nl. Wilhelm Wundt, Heinrich Holtzmann to Wilhelm Wundt, 28 December 1866.
146. UAL, Nl. Wilhelm Wundt, Adolf Hausrath to Wilhelm Wundt, 6 November 1904.
147. Geheimes Staatsarchiv Preußischer Kulturbesitz (hereafter GStA PK), VI. HA, Nl Althoff, No. 118, Otto Liebmann to Friedrich Althoff, 23 January 1884. See also Chap. 5 of this book.
148. GStA PK, VI. HA, Nl Althoff, No. 118, Carl Stumpf to Friedrich Althoff, 5 October 1893.
149. GStA PK, VI. HA, Nl Althoff, No. 118, Otto Hartwig to Friedrich Althoff, 27 October 1893.
150. UBL: BPL: 2389, Theodor Nöldeke to Michael Jan de Goeje, 7 December 1904. Nöldeke's emphasis. For earlier discussions between Nöldeke and De Goeje about Ahlwardt's work, see Christiaan Engberts, "The Scholar as Judge: A Contested Persona in Nineteenth-Century Orientalism," *BMGN—Low Countries Historical Review* 131, no. 4 (2016): 108.
151. Emil Behring to Otto Heubner, 29 June 1896, BNd: EvB/B 1/58.
152. Emil von Behring to Otto Heubner, 8 November 1905, BNd: EvB/B 1/59.
153. UAL, Nachlass Wilhelm Wundt, Gustav Theodor Fechner to Wilhelm Wundt, 13 April 1886.
154. UBL: BPL 2389, Michael Jan de Goeje to Theodor Nöldeke, 27 February 1870, 20 October 1905, 15 May 1906.

CHAPTER 3

The Editorial Balancing Act: Editors, Authors, Publishers, and Audiences

Scholarly Journals Before Peer Review

Today a respectable scholarly journal can easily be distinguished from other types of publications because most of its content has been subject to the thorough and double-blind evaluation process known as peer review. The history of peer review can be traced back to the seventeenth century. Both the Royal Society and de *Acádemie Royale des Sciences* established procedures to ensure that the books they published were in line with the printing prerogatives that their royal protectors had bestowed on them.[1] From 1665 onwards, these review practices were carried over to their newly founded journals, the *Philosophical Transactions* and the *Journal des Sçavans*. These evaluative processes, however, were quite rudimentary. The editors would make most decisions by themselves without consulting anyone else. The reliance on individual editors could give the journal a rather idiosyncratic character, which would also make it vulnerable to criticism. In the 1750s, the *Transactions* were mocked by an observer who drew attention to "trivial and downright foolish" contributions, such as "papers on a merman […], on the unicorn's horn, on the transmutation of water into maggots, and 'incontestable proof of a strange and surprising Fact, namely that Fish will live in Water.' "[2]

The minor scandal caused by this public ridicule persuaded the Royal Society to remove editorial responsibilities from the hands of the

© The Author(s), under exclusive license to Springer Nature Switzerland AG 2022

C. Engberts, *Scholarly Virtues in Nineteenth-Century Sciences and Humanities*, https://doi.org/10.1007/978-3-030-84566-7_3

individual editor-in-chief and appoint a so-called "Committee on Papers."[3] The members of this committee were also encouraged to seek the advice of outside experts if necessary.[4] In spite of these developments, the editor-centred model of journal management remained predominant in the nineteenth and early twentieth century. Most scholarly journals fashioned their editorial policies after those that were most common in newspaper publishing. Many of them also took the shape of official house organs of research institutes or scholarly societies and were therefore often simply managed by the director of the institute or society in question.[5] In this light, Christopher Booth convincingly argues that the history of scholarly journals "is very much bound up with the characters and personalities of the editors who, through the years, have contributed so much."[6] Modern-day practices of peer review defined as "the critical assessment of manuscripts submitted to journal by experts who are not part of the editorial staff" would only become common after the Second World War.[7] Stephen Lock points out that even in the 1960s, the editor of *Nature* still "relied mainly on expert opinions within the editorial office, taking the occasional article with him to ask a colleague's opinion over lunch at his London club."[8]

These developments in the anglophone world were also quite typical for nineteenth-century Germany. The pivotal role of individual editors is repeatedly stressed in Joachim Kirchner's historical survey of German periodicals.[9] The chemist Justus von Liebig's total control over the *Annalen der Chemie* is one of the best-known German examples of the editorial omnipotence of an authoritative and successful director of a research institute.[10] Another example from the same discipline is provided by Wilhelm Ostwald's *Zeitschrift für physikalische Chemie*.[11] The editorship of their own journal made it possible for Liebig and Ostwald to present their work to a broad academic audience and assured them of a podium for all the findings of their own and associated research groups.[12]

One of the journals analyzed in this chapter was managed in this way by a single editor who was also an institute director. Wilhelm Wundt was the only person responsible for the *Philosophische Studien*. The second periodical that I will discuss, the *Zeitschrift der Deutschen Morgenländischen Gesellschaft*, was the house organ of the *Deutsche Morgenländische Gesellschaft* (*DMG*), the German national society of orientalists. The statutes of the *DMG* set out that a board of editors would be responsible for the *Zeitschrift*. In practice, however, most editorial responsibilities tended to be assumed by the editor-in-chief. The *Literarisches Centralblatt für*

Deutschland, finally, was also managed by a single omnipotent editor, the Germanist Friedrich Zarncke.

It is interesting, though, to compare Zarncke's editorial challenges with those faced by Wundt and the editors of the *Zeitschrift*. Unlike Wundt, Zarncke could never fall back on contributions from his own research institute. And unlike the editors of the *Zeitschrift*, he did not have to take into account that his journal was published as the house organ of a national professional organization. Differences like these suggest that Lynn Nyhart is right in drawing attention to the fact that "journals have adopted quite a variety of practices expressing different relationships between editors, contributors and readers."[13] The similarities and differences between the *Studien*, the *Zeitschrift*, and the *Centralblatt* provide a fruitful ground for a further exploration of this variety. What is more, this variety of evaluative practices also provides a good starting point to reflect on collegial loyalty and critical independence.

Wundt's Early Editorial Experiences

The founding of the *Philosophische Studien* was a direct consequence of the high productivity of Wundt's laboratory. After its modest beginnings in the late 1870s, his institute for experimental psychology attracted an increasing number of ambitious young men. Between Wundt's arrival in Leipzig in 1875 and his death in 1920, more than 180 students wrote their dissertation under his supervision.[14] Most of these dissertations were based on experimental research in Wundt's Leipzig Institute for Experimental Psychology. The constant flow of experimental findings and papers called for a new platform to publish this rather coherent body of work. In 1880, Wundt complained about the lack of publication opportunities for his staff and students to his former pupil Emil Kraepelin: "At the moment I have some works […] that have been completed in my psychophysical laboratory [….]. I do not really know where they can be published. I would really like a periodical that offers a place where such works can be collected."[15] The *Studien* would eventually offer this place. This was not, however, Wundt's first experience as an editor.

Wundt had already been closely involved with the publication of the *Vierteljahrsschrift für Wissenschaftliche Philosophie* (Quarterly for Scientific Philosophy) in the 1870s. This journal was the initiative of the young philosopher Richard Avenarius. Wundt and Avenarius first met at Leipzig's *Akademisch-Philosophische Verein* (Academic-Philosophical Society) in

1875.[16] Avenarius shared Wundt's commitment to an experimental approach to philosophy. When he accepted Wundt's former Chair of Inductive Philosophy in Zürich in 1877, he founded his own journal. He asked his Leipzig acquaintance to support him. Wundt did not have any experience as an editor either, but he was eleven years Avenarius' senior, which put him in the role of a senior advisor. In this supporting capacity, he learned just how demanding—and even frustrating—the management of a new scholarly journal could be.

Some of the inconveniences were quite minor. During the first years of the *Vierteljahrsschrift*, Avenarius consulted Wundt on such everyday squabbles as pushy authors, whose unsolicited contributions were hard to reject.[17] Even if he could give Avenarius some advice about such minor issues, he could not do anything about the precarious financial situation of the journal and its limited resonance in academic circles. "It is truly embarrassing how little success the invested effort and costs have earned us," Avenarius complained in 1878.[18] Two years later, things had hardly improved: "[A]s little as our journal lacks in recognition and efficacy, so much does it still lack in a high enough number of subscribers."[19] Wundt, however, was not too impressed by these setbacks and for a moment he even considered publishing his own new journal as a supplement to the *Vierteljahrsschrift*.[20]

Not long after he first played with this idea, however, Wundt's enthusiasm for the *Vierteljahrsschrift* dampened. He complained to Kraepelin that Avenarius' journal did not offer enough space for publications from his laboratory.[21] One year later, in the summer of 1881, he criticized the *Vierteljahrsschrift* for having become "a bit too abstract and dry."[22] A quick look at the dissertations finished under Wundt's supervision shows why the need for a new platform sympathetic to experimental psychology was felt at this exact moment. The three dissertations completed in 1880 were clearly the product of research projects started before the psychological laboratory was established. Friedrich Wilhelm Horny had written a study of English utilitarianism, Paul Radestock defended a dissertation titled *Schlaf und Traum* (Sleep and Dream), and Otto Tischler wrote about East Prussian cemeteries (*Ostpreußische Gräberfelder*).[23] One year later, however, Max Friedrich was the first of Wundt's students to finish a dissertation based on experimental research.[24] In the same year, Martin Trautscholdt completed another dissertation on the basis of an experimental study in Wundt's laboratory.

In the German lands, it was quite common to publish such dissertations in journals before the nineteenth century.[25] The pages of the *Archiv für die Physiologie*, which was founded as late as 1795, were, for example, largely filled with reprints of dissertations.[26] By the time Wundt had founded his laboratory, however, this genre of academic publishing had largely vanished. Avenarius had persuasive reasons to refrain from printing dissertations in the *Vierteljahrsschrift*. He pointed out that they would not provide new content because they were published as books as well. He was also wary of the common expectation among new doctors that their work would be published on short notice. In December 1880, Avenarius explained to Wundt that he would only consider printing some of the very best dissertations if and only if their authors would forego payment and accept late publication.[27] Before the year was even over, he also retracted this modest promise: "At the moment, we have the experience that 'philosophical' essays in general, and philos[osphical] 'doctoral dissertations' in particular, hardly make up for the printing costs" because they tended to deal with topics "for which there is—alas!—not a lot of demand."[28]

Wundt realized that he could no longer count on a fruitful collaboration with Avenarius and got in touch with his long-time publisher Wilhelm Engelmann to discuss the possibility of establishing his own journal. Engelmann liked the idea but had quite outspoken ideas about it. Like Avenarius, he hesitated to renumerate the authors of dissertations: "I think that young people, students, should renounce all *material* benefit from their intellectual efforts."[29] He also proposed that the journal should not only print long experimental reports but also short announcements, book reviews, and theoretical reflections.[30] Wundt would personally contribute several theoretical papers but successfully resisted his publisher's pressure to add announcements and reviews. The *Studien* were largely published as Wundt had envisioned it when he first thought about starting his own journal: "I think it is best, at least form [...] to only publish standalone papers, no reading reports, reviews, etcetera."[31]

Wundt's apparent leverage in the negotiations with his publisher show the independence he enjoyed as an author and editor. This independence is further illustrated by the fact that the Engelmann publishing house accepted that the journal was hardly profitable. An overview of the costs and benefits by Emanuel Reinecke, who took over the management of the firm after Rudolf Engelmann's death in 1888, shows that the journal was not even close to turning a profit. Even though the first issue of the *Studien* had become profitable after having been available for more than seven years, none of the subsequent volumes had broken even.[32] Reinecke's proposed solutions to these financial setbacks were to either raise the sales price or limit the page count. He did not even contemplate, however, to ask Wundt to change the journal's character by moving away from its emphasis on the work of his doctoral candidates or by adding new features like book reviews and short announcements.

The independence that Rudolf Engelmann and Emanuel Reinecke allowed an editor of a hardly profitable journal is remarkable. After all, the Engelmann publishing house was a business without any special interest in promoting the work from Wundt's laboratory. The independence that Wundt enjoyed as an editor can easily be explained, however, by the closing paragraph of the long letter in which Rudolf Engelmann expressed his willingness to support Wundt's journal: "Finally, you will surely allow me to ask that you first turn to us when you are planning to publish any other more comprehensive study [...]. In the light of our pleasant personal relation, I would greatly appreciate it, if we would have a similar and enduring [professional] relationship."[33] Most of Wundt's books published by Engelmann went through several editions and therefore the publishing house was perfectly willing to allow its best-selling author a pet project as long as he would continue to publish his commercially more interesting works with them.

The considerations of Rudolf Engelmann illustrate that Wundt's editorial independence was not so much a feature of the emerging discipline of experimental psychology but primarily the result of the personal relationship between the publisher and his author, which was shaped by the high sales of the latter's books. Wundt's journal existed side by side with other types of journals publishing on similar topics. For example, Hermann Ebbinghaus' and Arthur König's *Zeitschrift für Psychologie*, founded in 1890, did not revolve around the research programs of its editors-in-chief in a similar fashion.[34] In 1903, Wundt eventually discontinued the *Philosophische Studien* and founded a new journal, the *Archiv für die Gesamte Psychologie*, with his former student, the pedagogist Ernst Meumann. After almost two decades of editorial omnipotence, however, he felt uncomfortable collaborating on an equal basis and publishing contributions about a wider range of topics. He felt particularly uneasy with Meumann's willingness to publish papers advocating the kind of practical approaches that were becoming increasingly dominant in pedagogy. He complained to another former student, Oswald Külpe, that it would be more appropriate to name the journal *Archiv für die Gesamte Pädagogik und Psychologie* instead.[35] In 1905, Wundt decided to put an end to his editorship at the *Archiv* and to publish a journal of his own again. From 1905 to 1917, he would manage a new journal, the *Psychologische Studien*. The *Archiv* continued without Wundt and would be published until the end of the Second World War.

Editing in Total Independence

Wundt used his editorial independence to turn the *Philosophische Studien* into the unofficial organ of his Institute for Experimental Philosophy. The twenty volumes published between 1881 and 1902 contained 324 contributions. Wundt wrote fifty-three of these. Most of the contributions were written by his closest colleagues. No less than 236 papers were written by scholars who had conducted research under his supervision. This group includes both doctoral candidates and laboratory assistants. The remaining contributions were either by pupils of Wundt's former students Emil Kraepelin and Oswald Külpe, Leipzig acquaintances like Gustav Theodor Fechner, and a miscellaneous group of people who cannot easily be categorized. This latter group includes such people as Harald Höffding, the Copenhagen mentor of Wundt's Danish student Alfred Lehmann. The combination of Wundt's editorial independence and the journals strict focus on Wundtian approaches to experimental psychology is most apparent in the way in which the *Studien* allowed Wundt to provide loyal support to those members of his circle who found themselves at the periphery of German academic life. Young men like Ludwig Lange, Julius Merkel, Gottlob Friedrich Lipps, and Friedrich Kiesow struggled for different reasons to establish an academic career after obtaining their doctorate with Wundt. Their publications in the *Studien*, however, ensured some sustained interest in their stubbornly continued research.

Lange was probably the most troubled of Wundt's doctoral students.[36] His father died in 1885, shortly after Lange had enrolled as a student in Leipzig. Without his father's continued support, it was doubtful whether Lange could even finish his studies. Wundt, however, recognized his mathematical talents and hired him as his assistant.[37] This position also allowed Lange to free up some time for his own research and write his dissertation. All this time, Wundt was well aware of his new assistant's vulnerability. Lange had informed him early on that he doubted "the health of his mental state" because he frequently experienced "agonising passive fantasies [and] obsessive thoughts."[38] During his first years under Wundt's supervision, however, he was very productive. He published his dissertation in two instalments in the *Studien* in 1886, received his doctorate in 1887, and even published some additional research papers in his *Doktorvater*'s journal.[39]

Shortly after receiving his doctorate, however, his life took a turn for the worse. In 1887, he suffered his first bout of mania. This was the start of a decades-long alternation between mania and severe depression. During his manic periods, he would be extremely talkative, prone to violent outbursts, and even inclined to leave his house wearing only a top hat and a waist coat. During his manic periods, he would just stay in bed for months and not talk to anyone.[40] A former colleague's account of an anonymous institute member who become "mentally deranged" is most likely a reference to Lange's troubles.[41] Of course, Lange could not hold on to his job as laboratory assistant. In December 1887, he wrote Wundt that his "chronical suffering of several years" had finally caught up with him and that he was no longer able to do his job.[42] Wundt, however, would continue to support him. He would publish his work in the *Studien* in 1888, 1894, and 1902.[43] In 1919, more than thirty years after Lange had left the Leipzig Institute, Wundt even—unsuccessfully—supported his application to a position as university librarian in Leipzig.[44]

Julius Merkel was another student of Wundt who failed to have an academic career. Merkel's dissertation was published in the 1883 issue of the *Studien*.[45] Even though Merkel would never find employment at a university, he continued to carry out research. Wundt not only published his former student's steady stream of research papers, but he also provided extensive feedback on his manuscripts. He encouraged Merkel to engage more with the studies of other doctoral researchers at his institute, such as Gustav Hermann Lorenz and Paul Adolf Starke.[46] He also discussed his doubts about some of Merkel's methodological choices, such as the reliability of measurements of the perception of the perceived loudness of sound.[47] Merkel would continue to be a regular contributor to the *Studien* until his final papers were published in its tenth volume in 1894.[48] A lot of his research was conducted and written after he had accepted a position as a high school mathematics teacher in a small town in the south-east of Saxony.[49] Even after Merkel had turned his back on academia, however, Wundt loyally continued to offer his former student critical feedback and a platform for publication.

Some of Wundt's students with a difficult career start eventually found permanent employment in academia. One of his most successful students with an unconventional career path was Friedrich Kiesow. When he started his studies with Wundt in 1891, he was already in his early thirties. He finished his dissertation three years later.[50] His relatively high age was the result of a combination of illness in his youth and limited family finances that forced him to work as a tutor for years.[51] The same illness was also the reason why he had never finished his *Abitur*, the official high school qualification for university admission. It was quite impressive that Kiesow had managed to be accepted to the University of Leipzig and even write a dissertation without *Abitur*. It was not possible, however, to submit a *Habilitation*, the official German requirement for an academic teaching position, without *Abitur*. Therefore, Kiesow left Leipzig for Turin to work with the Italian psychologist Angelo Mosso.

Even though Kiesow had left for Italy, his close relation with Wundt ensured his continued visibility in Germany. He kept sending manuscripts to his *Doktorvater* and he usually published them immediately in the *Studien*.[52] Wundt even asked his former student for a German translation of an article he had already published in Mosso's *Archives Italiennes de Biologie* because "the circle of readers of [the *Studien*] and Mosso's *Archiv* don't overlap anyway."[53] Wundt would publish Kiesow's work in the *Studien* until he finally acquired the *libera docenza* title, the Italian equivalent of *Privatdozent*, in 1899. Wundt enthusiastically wrote that he was glad to see that "two of [his] most diligent young assistants—you and Kirschmann—to whom the psychological teaching profession in Germany was barred [...] have found a suitable position abroad."[54] Once his Italian career had gained momentum, Kiesow would rely less often on his ties to Wundt. In 1919, he even followed his former teacher's example and founded his own journal, *Archivio italiano di psicologia*, to offer his own doctoral students a publication venue.[55]

A final example of the way in which Wundt's omnipotent editorship allowed him to provide loyal support to those former students who needed it the most can be found in his long-standing relationship with Gottlob Friedrich Lipps. After obtaining his doctorate with Wundt in 1888, Lipps wanted to start on his *Habilitation*. Because he was not financially independent, however, he needed a job to fund his research efforts. He preferred a position in a German university city because this would allow him to stay in touch with academic life while working on his *Habilitation*. For twelve long years, he worked as a high school teacher in Strasbourg and the nearby village of Hagenau. These demanding jobs did not, however, leave him enough time to work on his *Habilitation*. Therefore, after more than a decade of teaching in the Alsace, he approached Wundt. He asked him if he could help him find a job in Leipzig, which would allow him to work on his *Habilitation* in a familiar environment.[56] After a long search, Lipps finally found a job at a newly established Leipzig *Gymnasium* in 1902.[57]

Now that Lipps was back in Leipzig and doing research in his laboratory again, Wundt was well-positioned to help his former pupil. Lipps' *Habilitation* was approved as early as 1904. From this moment onward, Lipps would teach at his *alma mater* until he was appointed to Wundt's old chair for inductive philosophy in Zürich in 1911. Lipps' road from doctoral dissertation to professional appointment had been long and windy. An important reason why academic success had seemed attainable throughout decades of high school teaching in the Alsace was the fact that Wundt continued to publish Lipps' papers in the *Studien*. Between 1894 and the *Studien*'s dissolution, Wundt published nine papers by him.[58] Wundt held his former student's contributions in such high regard that he even published some of his papers without closely reading the manuscripts![59]

Wundt's management of the *Philosophische Studien* provides an elucidating example of the entanglement of independence and loyalty. On the one hand, the *Studien* draws attention to the way in which he had been able to achieve remarkable independence. The high sales numbers of his books convinced his publisher to allow him to manage a journal without worries about its potential readership and commercial viability. The growing number of students and researchers working at his Leipzig institute guaranteed that the journal did not depend on contributions by outsiders. For every issue, Wundt could freely select from his own work and a continuous flow of papers by his subordinates. As a full professor and laboratory director, it was also very easy for him to ask for changes and further elaborations to the submitted papers.

Wundt used this independent—or even omnipotent—editorial position not only to promote his own approaches to psychological and philosophical questions but also to provide loyal support to those former students who needed it the most. Of course, his loyal willingness to print his former students' work did not always pay off. It soon became clear that Ludwig Lange's mental problems made a successful university career nearly impossible and after a few years of high school teaching in a Saxon small town, it had become quite obvious that Julius Merkel would not continue to pursue an academic career either. Sometimes, however, Wundt's loyalty did pay off. He printed Friedrich Kiesow's papers in the *Studien* until he had overcome the disadvantage of a lacking *Abitur* and acquired a respectable academic position in Italy. Similarly, he published Gottlob Friedrich Lipps' contributions for decades until he finished his *Habilitation* and was appointed to a Zürich professoriate. These displays of loyalty were only possible from a position of editorial independence which allowed Wundt to support his less successful associates.

Fleischer: An Independent Team Player

Heinrich Leberecht Fleischer's editorship of the *Zeitschrift der Deutschen Morgenländsichen Gesellschaft* provides a very different example of the complex relation between critical independence and collegial loyalty. In 1846, the *Zeitschrift* was created as the organ of the *Deutsche Morgenländische Gesellschaft*—the German Oriental Society—and Fleischer was one of its founding editors. The commercial sustainability of the *Zeitschrift* was guaranteed by the fact it would be paid for by the financial contributions of the *DMG*'s membership. The search for a sympathetic publisher was facilitated by the fact that one of the journal's founders, the Sanskritist Hermann Brockhaus, was the son of Friedrich Arnold Brockhaus, the founder of the famous F.A. Brockhaus publishing house. The relationship between the *DMG* and the publisher of its journal was so close that from 1862 onwards O. Matthesius would be the societies treasurer on behalf of the publisher.[60] Even though Hermann Brockhaus obviously had the most intimate relation with the *Zeitschrift*'s publisher, Fleischer would turn out to be its most influential editor.

At the founding of the *Zeitschrift*, Fleischer was already one of Germany's foremost Arabists. He had studied Semitic languages during a period—the early nineteenth century—when these studies were somewhat underdeveloped in the German states. He had been one of the enterprising German young men who went to Paris to study with the world-famous French Arabist Antione Isaac Silvestre de Sacy.[61] From 1835 until his death in 1888, he would be professor of Oriental Languages in Leipzig and turn this city into the most important centre of German oriental studies. From the second half of the nineteenth century onwards, his students held professorial chairs across continental Europe.[62] The origins of the *DMG* can be traced back to a circle of like-minded oriental scholars that took shape around Fleischer in the early 1840s.[63] Therefore, Fleischer's central role in the management of the societies' journal was only to be expected.

The widespread acceptance of Fleischer's authority is reflected in Nöldeke's and De Goeje's sympathetic discussion of the major changes that he would make to manuscripts submitted for publication in the *Zeitschrift*. In a discussion of a paper by a shared acquaintance, Nöldeke makes the following observations: "[...] the shape of the essay will still change tremendously; I know how ruthless Fleischer is in deleting and correcting mistakes, how he often translates excerpts anew. [...] I can assure you that many articles in the *Zeitschrift* deserve Fleischer's name rather than that of the original author."[64] De Goeje would soon gain personal experience with Fleischer's interventionist approach to editing as well. Reflecting on the published version of a paper he had submitted, he noted that "Fleischer [had been] responsible for all those strange German words, which I would never have come up with myself."[65]

Nöldeke and De Goeje accepted such editorial inventions without complaining. Fleischer had one editorial habit, however, that they disapproved of and that was his tendency to print excerpts of private letters in the *Zeitschrift* without asking for permission. In early modern times, this practice was widely accepted. Editors with an extensive correspondence network were even expected to share all bits of news that could be of interest to their readers.[66] In the second half of the nineteenth century, however, this was no longer a standard practice. It bothered De Goeje, therefore, when he unexpectedly read excerpts of one of his private letters to Fleischer in the *Zeitschrift*. He desperately asked Nöldeke: "What can be done about this? When you write [Fleischer], do you write 'confidential' on letters that are not intended for a wider audience?"[67] He even shared his displeasure with Fleischer, who then pointed out that the publication of the letter could only be beneficial because it would prevent others from pursuing similar research projects in the future.[68] De Goeje was forced to accept—albeit grudgingly—that this was simply the way in which Fleischer balanced personal correspondence and editorial responsibilities.

Fleischer used his editorial clout not only to improve contributions and to share private correspondence. One of his most challenging tasks was to make sure that the views of all members of the *DMG* would be represented in its *Zeitschrift*. This was particularly complicated when a member's contribution evoked strong reactions. In such cases, a commitment to independent criticism encouraged the immediate publication of radical critiques. A commitment to collegial loyalty towards both individual peers and the *DMG* as a whole, however, suggested a less potentially divisive approach. Fleischer's attempts to balance criticism and loyalty are clearly visible in his correspondence with Nöldeke.

In 1879, for example, the editorial board of the *Zeitschrift* received a review copy of Friedrich Heinrich Dieterici's sloppy edition of a collection of fairy tales. Because he could not credibly publish a positive review, Fleischer tried to mitigate the impact that a less favourable review would have. He first wrote a letter to Dieterici in which he extensively set out his doubts about his book and prepared him for a less than glowing review. He then wrote Nöldeke to ask for a serious and rigorous review without any trace of vitriol or ridicule.[69] After Dieterici decided that he would immediately start working on a new, improved edition at his own expense, Nöldeke's mildly critical review was even entirely cancelled from the *Zeitschrift*.[70] Nöldeke would eventually write a favourable review of the improved edition for the *Centralblatt* in which he praised the importance of the work and the newly added list of corrections on the basis of Fleischer's earlier private comments.[71] In this way, Fleischer and Nöldeke were successful in balancing independent private criticism and collegial public loyalty.

Earlier that decade, Fleischer had already successfully dealt with a more delicate affair that involved a larger number of *DMG* members. The affair took place after the discovery of the Mesha inscription in the late 1860s. This discovery is still seen as "the greatest Biblical discovery of modern times" and provoked a great interest in artefacts from Old Testament times.[72] Soon the Jerusalem antique seller Moses Shapira seized the opportunity to show a large collection of earthenware and figurines to the German pastor Hermann Weser. Weser immediately got in touch with Konstantin Schlottmann, an Old Testament scholar from Halle, who was a *DMG* board member as well as a co-editor of its journal.[73] Schlottmann described the new discoveries in glowing terms in the *Zeitschrift*.[74] Because he was the only expert in this field on the *DMG* board, he was able to convince its general assembly to publish an advice to the Prussian Ministry of Education to purchase the whole collection.[75] The ministry immediate bought the artefacts and one year later they even bought a second collection from Shapira.[76] At the same time, however, more and more scholars started to express their doubts about the artefacts' authenticity.

The first high-profile critic was Albert Socin, who published his doubts in the *Zeitschrift* in 1873.[77] Fleischer was greatly concerned about the potential fallout of a polemic about this issue for at least two reasons. First, Schlottmann's initial enthusiasm had already resulted in the expensive purchase of two collections of artefacts by the Prussian government. Second, the division between advocates and critics of the artefacts' authenticity mirrored the main fault line within the board of the *DMG*. The society was jointly managed by scholars from Leipzig and Halle. Fleischer anticipated that most criticism would come from scholars with ties to Leipzig, while their Halle peers would continue to support Schlottmann's insights. He feared that such partisan dissent between Leipzig and Halle could trigger a "schism between Leipzig and Halle" that would make the *DMG* unmanageable.[78] Therefore, he urged Socin to postpone further criticism until Schlottmann would have published a more extensive illustrated study of the artefacts.[79] He also tried to ensure that all criticism of Schlottmann would be published in the *Zeitschrift* because this would allow him to ensure that the debate would be even-handed and polite.[80] Fleischer was largely successful. Most of the early criticism of Schlottmann's assessment of the artefacts was written by foreign scholars and published abroad.[81] The discussion in Germany only continued when Schlottmann used the *Zeitschrift* to defend himself against the criticism of his French detractor Charles Clermont-Ganneau.[82]

This opened the floodgates to a stream of critical discussions of the authenticity of the artefacts in Germany as well.[83] Socin published a book-length rebuttal of Schlottmann with his friend and colleague Emil Kautzsch.[84] Nöldeke joined the discussion with a long essay in the *Deutsche Rundschau*, a book review in the *Centralblatt*, and with a warning against Shapira's business practices in a review of Socin's recently published Baedeker travel guide for Palestine and Syria.[85] So even though Fleischer was able to delay the discussion, he was neither able to prevent it nor was he able to ensure that it would take place on the pages of the *DMG*'s *Zeitschrift*. His repeated calls for a civil debate, however, were more successful: "[W]ith God's help, the case […] has solved itself. The 'God' that helped us with this is mainly the spirit of moderation in thought and expression of thought, for which I have to praise the opponents Schlottmann and Kautzsch [… who] focused on an honest debate, which is unimaginable without mutual respect and avoiding all personal remarks and offensive insinuations."[86]

August Fischer's Fragile Authority

Fleischer's authority as editor of the *Zeitschrift* was largely based on the fact that he was widely recognized as the *eminence grise* of German orientalism. Prominent scholars like Nöldeke and De Goeje even referred to him as "*der gute Alte*" or "*Papa Fleischer.*"[87] Because none of his co-editors and successors had a similar reputation, they were much more dependent on the sympathetic cooperation of their peers. This is particularly clear in the editorial career of August Fischer, who became the *Zeitschrift*'s editor-in-chief in 1903. During his editorial tenure, Fischer grappled to find a balance between assuring a civil debate and giving room to a wide variety of insights. In addition, he also struggled to find a balance between his role as an impartial journal editor and his desire to contribute to debates in his own field of expertise.

In editing manuscripts, Fischer took the same liberties as Fleischer: he deleted expressions that he considered to be inappropriate and corrected mistakes.[88] Unlike Fleischer, however, Fischer was publicly asked to account for his changes. In 1905, the executive board of the *DMG* received a complaint from Gustav Jahn, an emeritus professor of Semitic Languages in Königsberg.[89] His manuscript about the Mesha inscription had been corrected by Fischer and he angrily enquired whether the charter of the society even allowed for such unsolicited changes.[90] Luckily for Fischer, nobody on the board thought very highly of Jahn. Board member Franz Praetorius bluntly stated that "it is generally known that Jahn's intellectual powers haven't been normal for quite some time" and that "his scholarly works have slipped more and more into the domain of the ridiculous."[91] Fischer further defended his edits by pointing out that most of them consisted of deleting or softening personal attacks.[92] All other changes had been "so minor that only a troublemaker or a lunatic" would mind them.[93]

Ernst Windisch, another board member with long editorial experience, also stood up for Fischer. He argued that it was the duty of the *Zeitschrift*'s editor-in-chief to make sure that no member of the society would be provoked without good reason.[94] Therefore, if Fischer would not have edited Jahn's vicious polemic, the only other option would have been to flat out reject it. Rejection had not been an option, however, because Jahn's contribution had been a reaction to a recent article by Eduard König and the editors wanted to allow different points of view on the pages of the *Zeitschrift*. If Jahn's polemic had not been a reaction to another article, Fischer argued, he would have preferred to not publish it at all: "I have

[…] regretted the inclusion of König's essay for a long time because it means that I have to let [Jahn], who does not write anything that is not complete or half rubbish, have his say in the *ZDMG* as well." The board therefore decided that Jahn would not receive any form of satisfaction. It even allowed König one more reply "for reasons of fairness" with an added footnote that stated that this would be the end of the discussion.[95]

Fischer's lack of paternal authority became an even bigger problem when he publicly joined scholarly debates. Soon after his appointment as editor-in-chief, he introduced a new section to the *Zeitschrift*, the *Redakteurglossen* (editorial comments), which consisted of "short remarks with critical or complementary content in a casual way, as they came to me while reading the essays and announcements that I received for the *Zeitschrift*."[96] He added that he hoped that nobody would take offense. However, his commentary soon turned into a prolonged debate with the Berlin Semitist Jakob Barth. This debate would play out on the pages of the *Zeitschrift*, in the private correspondence between Barth and Fischer, and in the correspondences of other influential members of the *DMG*. The dispute turned into a minor scandal after Barth had voiced his indignation to Nöldeke, with whom he had studied, and De Goeje, on whose al-Ṭabarī edition he had worked.

Nöldeke and De Goeje were not pleased with Fischer's *Redakteurglossen*. In letters to De Goeje and the Hungarian orientalist Ignaz Goldziher, Nöldeke stated that in his capacity as editor-in-chief, Fischer did not have the right to place himself above the *Zeitschrift*'s contributors in such a pedantic way.[97] De Goeje believed that Fisher's decision to publish his commentary must have been the result of bad advice.[98] When Fischer realized his *Redakteurglossen* had caused a lot of commotion, he wrote De Goeje to ask him whether he thought that his comments had damaged the cause of the *DMG*.[99] De Goeje's reply has been lost but must have been harsher than Fischer had expected. In his next letter to De Goeje, he desperately tried to defend himself against his Leiden colleague's assertions that had presented himself as a "chief justice" in front of whom "no appeal is possible."[100] Even though the executive board of the *DMG* was not as harsh as De Goeje, Fischer decided to discontinue his *Redakteurglossen*.[101]

This decision did not, however, end the tension between Barth and Fischer. In 1907, Barth published a book in which he criticized Fischer's "intemperate outbursts" and even added the rather extraordinary charge that he had "concealed scientific facts."[102] Understandably, Fischer was not amused. He used his position as the *Zeitschrift*'s editor-in-chief to put

Barth back in his place with snide remarks such as "With excessive confidence alone, one cannot be successful in scholarship in the long run, at least not in Arabic studies."[103] Now it was Barth's turn again to be indignant and submit an angry rebuttal of Fischer for publication in the *Zeitschrift*. Unsurprisingly, Fischer was not too keen on publishing Barth's diatribes and he put the case before the board of the *DMG*.

Fischer made his dislike of Barth's reply very clear: "Everywhere in modern society an inflicted injustice or insult is amended either by taking it back or by calling the offender to order, but not by giving the offended the right to insult the offender [...] to the best of his abilities."[104] Franz Praetorius defended Fischer's point of view as vehemently as he had done during the Jahn affair. He argued that the "sad history would only become sadder with the publication of an upset [...] *ad hominem* reply by Barth."[105] While almost all board members agreed with Praetorius, Nöldeke disagreed. He argued that Fischer's attempts to deny Barth a rebuttal were so questionable that he should resign his editorship.[106] When he recounted the board's discussion to De Goeje one day later, he mentioned that "no matter how the struggle between Fischer and Barth may turn out, it is certain that Fischer will give up his editorship. And that is a good thing!"[107]

Nöldeke's prediction was wrong: Barth's reply was never printed, and Fischer only published a short, half-hearted apology and continued to be the *Zeitschrift*'s editor-in-chief.[108] Nöldeke indignantly decided that he would not contribute to the *Zeitschrift* anymore as long as Fischer stayed in place.[109] Barth even revoked his membership of the *DMG*.[110] At the general assembly, the Arabist Carl Heinrich Becker presented an open letter in which he objected to Fischer's treatment of Barth. The letter was co-signed by twenty-five prominent members of the society, like Nöldeke, Goldziher, Carl Bezold, and Christiaan Snouck Hurgronje.[111] Fischer's days as editor-in-chief were now numbered. He had already considered resigning during the early days of his dispute with Barth in 1905.[112] When his staunchest defender, Praetorius, left the *DMG*'s board in 1910, Fischer threw in the towel as well.[113] Later that year, Barth re-joined the organization and Nöldeke published in the *Zeitschrift* again for the first time since 1908.[114]

The examples of Fleischer and Fischer show that Wundt's independence at the *Studien* was not a self-evident feature of the editorship of a nineteenth-century scientific journal. Only because he was a well-respected authority in his field, Fleischer was able to take certain liberties, such as correcting and embellishing submitted manuscripts, and publishing private letters. Fischer's decisions, however, were constantly challenged. The resistance against his editorial choices can partly be understood in the light of his lacking personal authority. To add insult to injury, he was more willing than Fleischer to get personally involved in scholarly disputes. He found it difficult to find a balance between independent criticism of his peers and the expectation that—as editor-in-chief—he would stand above all parties. This weakened his position as the *Zeitschrift*'s editor and eventually was the reason for his resignation.

The experiences of Fleischer and Fischer also illustrate how considerations of loyalty extended beyond their relations to individual scholars. The *DMG* provided a new focus of loyalty, both to the organization itself and to its diverse membership. The ensuing need to treat all members equally encouraged the development of two editorial policies. Fleischer was particularly successful in realizing the first policy: making sure that the *Zeitschrift* would predominantly print sympathetic reviews of the works of the society's members. However, both Fleischer and Fischer occasionally had to resort to a second policy: ensuring that somebody who had been criticized on the *Zeitschrift*'s pages would at least be able to publish a reply. Putting these policies into practice can be understood as exercising loyalty both towards the *DMG* and to the diverse group of scholars that made up its membership.

A Review Journal in a Commercial Environment

Wundt, Fleischer, and Fischer shared one fortunate circumstance: they did not have to worry about money. The continued existence of the *Studien* was guaranteed by the sale of Wundt's books and the *Zeitschrift* was financed by the *DMG*'s members and published by the family of one of its founding members. One of the main differences between these journals and the *Literarisches Centralblatt für Deutschland*, was that the latter was a commercial publication that needed to make a profit to survive. Therefore, its editor's opportunities for independent decision-making were clearly more limited than those of his colleagues at the *Zeitschrift*. These opportunities were further limited by the fact that the *Centralblatt* covered a wide range of topics and could not fall back on the members of its own professional society—like the *Zeitschrift*—or the staff of its own laboratory—like the *Studien*.

The *Centralblatt* was founded as a weekly review journal in 1850, after similar journals like the *Literarische Zeitung* and the *Jenaische Allgemeine*

Literatur-Zeitung were discontinued after the political upheavals of 1848.[115] The new journal was founded by the publisher Georg Wigand, the classicist Otto Jahn, and the legal scholar and historian Theodor Mommsen.[116] Friedrich Zarncke was one of the first collaborators and the first issue of the *Centralblatt* already mentions him as editor. In the aftermath of the 1848–1849 events, the University of Leipzig fired Jahn and Mommsen and the full editorial responsibility for the journal fell on Zarncke.[117] Wigand, who moved in the same liberal circles as Jahn and Mommsen, decided to sell the journal. From 1852 onwards, it would be published by the publishing house Avenarius & Mendelssohn. From 1855 onwards, Eduard Avenarius, the father of Richard Avenarius, would be its sole publisher.[118]

Even though Eduard Avenarius had not attended university, he was well-connected to the Leipzig intellectual circles. He even had family ties to the Brockhaus family: he was married to Cäcilie Geyer, the half-sister of the wife of Hermann Brockhaus' brother Friedrich.[119] Avenarius' son and family chronicler Ludwig describes his father as a publisher who did not really care for the *Centralblatt*'s profitability: "[F]or him it was less about rich monetary profit than about serving the public good."[120] However, even though Avenarius may have been an exceptionally sympathetic publisher, he could not afford to work at a loss. His correspondence with Zarncke shows that the publisher and the editor were constantly discussing ways in which they might be able to reach a wider audience. Hoping that its inclusion would increase advertisement revenues, Avenarius, for example, convinced Zarncke to add a regular overview of the courses offered at various German universities.[121]

To increase their readership among university researchers, Avenarius also repeatedly suggested to report on the latest academic news. His initial idea was to simply list "promotions and deaths."[122] Two years later, in the spring of 1864, he also proposed to publish what "from certain sides could be interpreted as gossip" because it does not hurt to throw "a pike in the carp pool" every once in a while.[123] Two weeks after Avenarius brought up this idea, Zarncke published the first *Personalnachrichten* in the *Centralblatt*. However, their matter-of-fact tone was closer to Avenarius' earlier ideas about "promotions and deaths" than to his later, more sensationalist, suggestion. Zarncke clearly did not feel obligated to follow up on all the whims of his publisher. This is further illustrated by his refusal to add a section called 'Questions to the scholarly world' and his rejection of

Avenarius' idea to publish reviews of "the most outstanding publications in belletristic literature" as well.[124]

Even though the *Centralblatt* never turned into a great source of profit, the journal's continued existence was not under thread until the 1870s. In 1874, however, a new review journal was founded in the nearby university town of Jena, the *Jenaer Literaturzeitung*. The *Literaturzeitung* was managed by Anton Klette, an experienced editor who had earlier led the *Rheinische Museum für Philologie* and enthusiastically embraced the opportunity to edit a journal that aimed at a wider audience.[125] Like the *Centralblatt*, the *Literaturzeitung* appeared on a weekly basis. Avenarius immediately recognized the threat it posed: in March 1874, he referred to it as "our rival" and emphasized the importance of keeping track of "the competition from Jena."[126]

Avenarius made a detailed analysis of the similarities and differences between the *Centralblatt* and the *Literaturzeitung*. He calculated that in its first eleven issues of 1874, the *Centralblatt* had reviewed 235 books, while the *Literaturzeitung* had only discussed 166. He decided against advertising his own journal's broader selection, however. This would give his competitors the opportunity to point out that they, indeed, reviewed "fewer book, but of course all the important ones, and […] more extensively."[127] Still, Avenarius continued to count reviews. In early 1876, he informed Zarncke that the *Centralblatt* had reviewed 1131 works in 1874 and 1199 books in 1875, while the *Literaturzeitung* had reviewed only 789 and 815, respectively.[128] These numbers were not a reason for great confidence for the future, however. The high number of reviews was partly caused by a reduced number of advertisements, which had forced Zarncke to push for more reviews than before.

One response to this worrying development was an appeal to the loyalty of the frequent contributors to the *Centralblatt*. Avenarius carefully calculated that at least ninety-five contributors to his journal had also written reviews for the *Literaturzeitung*. He told Zarncke that he was worried that they would lose indispensable contributors to their Jena rival. After all, it was unlikely that anyone would want to review the same book for two different journals.[129] Zarncke's correspondence with Wilhelm Wundt shows that he immediately acted on his publisher's worries. Throughout the 1870s, Wundt had been a regular contributor to the *Centralblatt*. Already in 1874, he had expressed his willingness to contribute to the *Literaturzeitung* as well. In 1875, he was finally asked to review a large pile of books for the Jena journal, which was immediately noticed by

Avenarius and Zarncke. Wundt confirmed that he had indeed occasionally received the same books from Leipzig and Jena but reassured Zarncke that in those cases he had only written a review for the *Centralblatt*.[130] Wundt would prove his loyalty to Zarncke in the following years. In 1876 and 1877, he would write thirty-eight reviews for the *Centralblatt* and only five for the *Literaturzeitung*. After 1877, he stopped writing for the rival from Jena altogether.[131]

Another consideration in dealing with the competition from Jena was the anonymity of reviews. During the first quarter century of its existence, the *Centralblatt* either printed its reviews anonymously or signed with a short cipher. During the 1860s, for example, Nöldeke often signed with a cross resembling the letter "X."[132] Anonymous reviewing was a common practice outside of Germany, as well. In British journals, for example, reviews were usually published anonymously, which created "a sense of the author as a neutral, all-seeing guide, free from human subjectivity."[133] Zarncke aimed at a similar objectivity through anonymity. In an essay celebrating the twenty-five-year existence of the *Centralblatt*, he argued that anonymous reviews encouraged substantive debate rather than personal recriminations. In a not-quite-modest tone, he added that it could have come across as distasteful self-advertising if the *Centralblatt* would have flaunted its impressive roster of contributors.[134]

The practice of anonymous reviewing became somewhat old-fashioned in the 1870s. In Britain, the "first completely signed periodical of the century" was founded in 1877.[135] Avenarius soon found out that the *Literaturzeiting* was widely praised for only publishing signed reviews.[136] This enthusiastic reception could not have come as a complete surprise to Avenarius. As early as 1851, Zarncke had received complaints about anonymous reviews. Nöldeke's teacher Heinrich Ewald, for example, explained that it was impossible for him to assess the merit of a review if he did not know who its author was.[137] Avenarius and Zarncke had ignored such complaints for twenty-five years but could not continue to do so in the light of the enthusiastic reactions to the *Literaturzeitung*'s editorial policy. From 1874 onwards, they encouraged all reviewers to sign with an easily recognizable cipher.[138] During the second half of the 1870s, Nöldeke would sign almost all his reviews with "Th.N." and "W.W." would be Wundt's *nom de plume*.

In the essay that opened the 1874 anniversary issue of the *Centralblatt*, Zarncke hoped to escalate the competition between his journal and the *Literaturzeitung*. Avenarius, on the other hand, was not sympathetic to the idea of a potentially prolonged editorial polemic and urged Zarncke to mitigate those parts of his outline of the *Centralblatt*'s history that could be interpreted as "open or covert attacks" of their peers in Jena.[139] Zarncke reluctantly complied with his publisher's suggestions. He deleted a reference to other scholars' "detrimental ambition" and replaced the words "hack writers of the lowest rank" with the slightly less offensive "hack writers."[140] In addition, his statement that authors whose work was not "virtuous" enough did not deserve the opportunity to reply to critical reviews was replaced with a somewhat less harsh comment about work that was not "solid" enough to merit a defence.[141] Avenarius eventually printed additional copies of the anniversary issue and sent them to "the most important newspapers" in Germany in the hope that it could generate some badly needed free publicity.[142]

In the end, the *Centralblatt* won the competition with the *Literaturzeitung*. It is very unlikely, however, that the anniversary essay, the signed reviews, and the pressure on authors who contributed to both journals were responsible for this success. The downward trajectory of Anton Klette's career was the main reason for the *Literaturzeitung*'s demise. In June 1878, the senate of the University of Jena forced him to resign for neglecting his responsibilities as a librarian.[143] He continued to work for the *Literaturzeitung* from Magdeburg for one more year. In 1879, however, the journal printed an announcement that informed its readers that the journal would be discontinued because its editor had suddenly disappeared.[144] Hardly anything is known about Klette's life after 1879: it has only been reported that he disappeared without a trace in the United States after either 1879 or 1896.[145] Even if their own actions might not have been very consequential, Avenarius and Zarncke were more self-assured from 1879 onwards. When a new review journal was launched in 1880, the *Deutsche Literaturzeitung*, they confidently stated that they were "completely at ease with this new competitor."[146]

Zarncke's Continuous Balancing Act

His own publisher was not the only stakeholder of the *Centralblatt* that Zarncke had to appease. The publishers of reviewed books had an obvious interest in favourable reviews of their catalogue. Already in the 1860s, Avenarius warned Zarncke that publishers might be less likely to send free review copies if they could only expect critical assessments.[147] The correspondence between F.A. Brockhaus and Zarncke shows that publishers did not shy away from protesting negative reviews. After pressure from the publishing house, for example, Zarncke printed Camillo Kellner's reply to a harsh assessment of his *Kurze Elementargrammatik der Sanskrit-Sprache*.[148] Because he had to protect his relationship with his contributors as well, Zarncke did not always give in to disappointed publishers. When F.A. Brockhaus objected against a review of a recently published study by H.E. Hergt, Zarncke consulted the book's critical reviewer before answering the publisher.[149] After the Brockhaus representatives had read the reviewer's further critical elaborations, they accepted that they just had to "acquiesce to what had been said about [the] publication in the *Centralblatt*."[150]

Zarncke had to be careful to maintain a careful balance between the *Centralblatt* and the book publishers. On the one hand, the journal depended on the cooperation of the publishers. If they would refuse to send review copies, Zarncke would have nothing to review. Therefore, he had good reason to refrain from publishing too many unfavourable reviews. On the other hand, Zarncke had some space for negotiation as well. Even if the publishers were wary of critical reviews, they realized that favourable reviews could be an asset—and they depended on editors like Zarncke to print such reviews. In addition, the *Centralblatt* was not completely dependent on free review copies either. Zarncke was frequently approached with spontaneous offers to review books already owned by the reviewer.[151] Even though these reviews tended to be favourable because the authors had often acquired the book for a good reason, such offers put at least some limit to the extent to which the *Centralblatt* depended on publishers.

In addition, Zarncke's network was valuable to publishers as well. In his capacity as editor, Zarncke had established a working relationship with specialists in almost all fields. F.A. Brockhaus and other publishers realized that good relations with such well-connected people could be assets. In 1869, for example, F.A. Brockhaus asked Zarncke if he could find someone who would be able to provide an expert opinion about August Knötel's recently submitted manuscripts about the "Lycian Trojans."[152] Zarncke consulted an expert in his network and two weeks later F.A. Brockhaus thanked him for confirming the doubts they already harboured.[153] The book was never published. In the following years, Brockhaus would also ask for Zarncke's mediation to solicit an expert opinion on Georg August Pritzel's *Thesaurus literaturae botanicae* and to help find contributors to the firm's famous *Konversationslexicon*.[154]

Like their publishers, the authors of book reviewed in the *Centralblatt* had a stake in a favourable review of their works, as well. Unlike them, however, they did not have the clout to pressure Zarncke into keeping harsh criticism out of the journal's pages. After all, the business model of the *Centralblatt* did not in any way depend on the collaboration of reviewed authors. Zarncke did consider it to be "a commandment of duty and honour," however, to allow a response to authors who felt that they had been treated unfairly.[155] All the same, he realized that this could be the starting point of a prolonged debate, that would be "of no use for scholarship" and "could only be unpleasant for the reader."[156] He accepted that he should allow the reviewer to reply to a reviewee's response but decided to cut off any further debate. He therefore ended up with an approach that was similar to Fischer's policy at the *Zeitschrift*.[157] The opportunity to engage in such a short-form debate was used quite frequently. Between 1850 and 1891—the year of Zarncke's death—179 responses were published and 151 of these replies were printed side by side with a reaction of the reviewer.[158]

Zarncke's willingness to publish responses should be understood in the light of his need for a large group of collaborators. Every year, he had to fill fifty-two issues of the *Centralblatt* and he could not afford to damage his good relations with frequent contributors like Nöldeke and Wundt. Between 1871 and 1880, the former contributed 96 reviews and the latter 123.[159] Zarncke had to be tolerant towards such valuable collaborators. He decided, for example, to hold back in his uneasy relationship with Nöldeke, who he personally disliked. After a minor dispute with Zarncke, Nöldeke temporarily stopped contributing to the *Centralblatt*.[160] A letter by Nöldeke's colleague and frequent *Centralblatt* contributor Alfred Gutschmid illustrates Zarncke's problems with Nöldeke: "Once you get to know him personally, you will notice that he can be rude and you will find him informal."[161] In this same letter to Zarncke, however, Gutschmid praised Nöldeke's lack of "scholarly obscurity" and expressed his relief about the fact that his colleagues had made peace again. In the end, Zarncke's dependence on frequent contributors outweighed his personal dislikes.

This dependence influenced not only *who* could publish in the *Centralblatt*, it also had an impact on *what* could be published. Regular contributors, for example, occasionally convinced Zarncke to publish anonymous reviews of their own work! In 1872, Franz Delitzsch submitted a review of a text edition he had just published with Seligman Baer.[162]

Not only did Zarncke accept the review, but he even asked Delitzsch for another review of his own work two years later.[163] On another occasion, Delitzsch did not submit his own review but suggested an appropriate reviewer instead. When he published a new text edition with Baer, he got in touch with Nöldeke to discuss the most suitable reviewer. In the end, they settled on Nöldeke himself.[164] Wundt recognized this as a very common practice. After reading a particularly favourable review in the *Centralblatt*, he argued that this was an obvious "literary token of friendship, as they occur so often in the field of criticism."[165]

Zarncke allowed frequent reviewers some liberties regarding the length of their reviews, as well. He generally preferred short reviews. Most reviews were not longer than one page and many were even shorter. This allowed them to review more books than others, such as the *Jenaer Literaturzeitung*.[166] Some reviewers, however, preferred to submit longer essays. When a regular contributor would submit a lengthy review, Zarncke usually set aside his preferred policy and published it anyway. Richard Avenarius' four-page discussion of Heymann Steinthal's *Einleitung in die Psychologie der Sprachwissenschaft* was immediately published.[167] That this was not only done because he was the publisher's son is illustrated by a four-page review by Nöldeke's former student Georg Hoffmann. His favourable assessment of his teacher's new Syriac grammar was also published without any delay.[168]

The correspondence between Zarncke and the aforementioned Alfred Gutschmid shows that Zarncke sometimes even allowed the publication of questionable content. In 1873, he asked Gutschmid to review *Geschichte Bochara's oder Transoxanien* by the controversial Hungarian Jewish scholar Hermann Vámbéry. Because of his somewhat sensationalist travel accounts, Vámbéry was not taken very seriously by his German colleagues.[169] His most acclaimed student, Ignaz Goldziher, remembered his lectures as full of "self-praise, bragging, presumptuous appraisal of his own achievements" and "harmful dilettantism."[170] Surprisingly, the early reviews of his *Geschichte Bochara's* had been quite favourable. Gutschmid did not agree with these reviews in what he called "semi-scholarly journals" and set out to put Vámbéry back into his place.[171]

In his correspondence with Zarncke, Gutschmid not only criticized Vámbéry's scholarship but also drew attention to his Jewishness. He referred to him as "Hirsch Bamberger," an easily recognizable Jewish name. He added some more explicit slurs as well, arguing that it was necessary to "shut the Jewish boy's mouth." Vámbéry's new book was obviously nothing more than the "puffery of an impertinent Jewish boy." Gutschmid realized that the *Zeitschrift* of the *DMG* would be the most appropriate place for a long, detailed review of Vámbéry's work but argued—probably correctly—that Fleischer would not accept his polemic essay. With obvious frustration, he referred to the "idyllic still life that Fleischer has established among German orientalists."

Zarncke immediately published the harsh assessment that accompanied Gutschmid's letter in the *Centralblatt*. Gutschmid had used it to scold Vámbéry's "total lack of knowledge," argued that he wrote "like a blind man about colour," lambasted his "circular reasoning and other offences against logic," characterized his book as "abysmal fibbing," and posed the rhetorical question "what forced the author to undertake something for which he lacked […] both the intellectual and external conditions." He also criticized Vámbéry's writing style as bombastic and "taking the fateful step from the lofty to the ridiculous." In his last paragraphs, Gutschmid hinted, if not at Vámbéry's Jewishness, at least at the fact that he was not German by criticizing the crudeness of his language and by drawing attention to some incorrect grammatical constructions.[172]

As was to be expected, Vámbéry was offended by Gutschmid's reproaches. In his response in the *Centralblatt*, he defended his scholarly trustworthiness and admitted that as a Hungarian speaking several Western and Eastern languages, it was almost impossible not to make some minor mistakes in German.[173] Gutschmid used Vámbéry's remarks about his linguistic proficiency to subtly draw attention to his Jewishness, after all: "[T]he only thing I knew thus far about the author of Bochara, is that he is actually called Bamberger […] no wonder that I could only assume that German, or at least not Hungarian, was his mother tongue."[174] Zarncke's willingness to let this antisemitic dog whistle pass shows just how far he occasionally went to please his regular contributors. At the same time, the incident also shows that he also drew some lines. After reading the manuscript of Vámbéry's response to his review, Gutschmid pointed out some of his spelling mistakes to Zarncke with a request not to correct these. This would expose Vámbéry's non-German background even more clearly.[175] This time, however, Zarncke did not give in: misspelled words like "*Erwiederung*," "*pracktischen*," and "*Kritick*" were correctly printed as "*Erwiderung*," "*praktischen*," and "*Kritik*."

Editors, Publishers, Authors, and Audiences

The case studies in this chapter provide an in-depth look at the ways in which collegial loyalty and critical independence shaped the practice of editing scholarly journals in a time before the ubiquity of modern-day double-blind peer review. Because decisions about the acceptance of manuscripts—with or without minor or major changes—could not be made on the basis of external assessments, the individual qualities and considerations of the editor-in-chief were even more pivotal in the late nineteenth century than they are today. The editors-in-chief were involved in a continuous effort to balance the interests of publishers, contributors, and their readers. The character of a scholarly journal depended both on the editor's ability to satisfy this diverse group of stakeholders and on his ability to maintain some form of independence from them. Different editors with different aims developed different strategies to balance their objectives with the expectations of the other involved parties.

The interest of publishers in these journals is rather obvious. They carried the financial risk of the endeavour and hoped to make a profit. For some publishers, this was sufficient reason to become involved with the content of their journals. Avenarius pitched ideas to Zarncke with varying degrees of success. Engelmann half-heartedly and unsuccessfully tried to convince Wundt to add announcements and book reviews to the *Studien*. The firm F.A. Brockhaus, finally, did not interfere with the *Zeitschrift*'s editorial choices at all. The extent to which editors could manage their journals independent from their publishers' meddling can at least partly be explained by the journals' financial positions. The fact that the *Zeitschrift* was the *DMG*'s official journal guaranteed a paying readership of about 500 members.[176] Engelmann allowed Wundt all the independence he desired because his books paid for it. Publishing the *Studien* for little to no profit was a sound investment because it was a way to accommodate one of the publisher's best-selling authors.

Avenarius was more involved with the journal he published because the *Centralblatt* was not a guaranteed cash cow. Its readership was not as reliable as that of the *Zeitschrift* and Zarncke was neither a best-selling author nor willing to publish all his books with Avenarius. Especially when the financial position of the *Centralblatt* was threatened by the founding of the *Jenaer Literaturzeitung*, Avenarius pushed for changes, such as the introduction of a *Personalnachrichten* section and the end of anonymous reviewing. Even though Zarncke successfully pushed against turning the

Personalnachrichten section into an academic gossip column, he could no longer manage the journal exclusively on the basis of his own independent judgement. He was forced to accept at least some of his publisher's proposals.

The perceived relation between the journal and its audience also decisively shaped editorial practices. Wundt never seems to have cared too much about the question whether his journal reached a broad audience. After all, the journal was primarily established to provide a loyal service to his friends, former students, and collaborators. The *Studien* was the only journal that offered an easily accessible platform for dissertations written under Wundt's supervision, as well as for research papers of some of his former students, who found themselves at the periphery of the academic world.

The editors of the *Zeitschrift* found themselves in a very different situation. Because it was the official journal of the national organization for German orientalists, its rather broad readership and its pool of contributors largely overlapped. Because they served a diverse group of *DMG* members, the *Zeitschrift*'s editors had to walk a fine line between publishing different points of view and preventing prolonged and potentially heated debates. Especially Fischer found that it was almost impossible to satisfy everybody and became involved in protracted complaints procedures and polemics. "Papa" Fleischer's authority was more widely accepted, which made it easier to manage these issues. This was not appreciated by all member of the *DMG*, however. Deploring what he called the "idyllic still life that Fleischer has established among German Orientalists," Alfred Gutschmid took his harshest assessments of his colleagues elsewhere.

Avenarius and Zarncke, finally, managed a journal that aimed at a broad audience with only a minor overlap between contributors and readership. Because of the aforementioned occasionally precarious financial position of the *Centralblatt*, the perceived preferences of its readers guided editorial decisions. The *Personalnachrichten* and the attempt to limit anonymous reviewing were inspired by the readership's supposed preferences. In addition, Avenarius successfully argued against the inclusion of overviews of technical and agricultural periodicals, because he thought that there was no overlap between the potential readers of the *Centralblatt* and people interested in these topics.[177]

Even though the editors of the *Studien*, the *Zeitschrift*, and the *Centralblatt* had very different relations to their publishers, readership and contributors, their editorial practices have one remarkable feature in common. They all had their own reasons to limit the opportunities for debate. The shared academic background of almost all contributors to the *Studien* prohibited the variety of opinions that could have been the starting point of a prolonged dispute. After all, each contribution built in some way on Wundt's ideas about experimental psychology and Wundt himself decided whether a paper would be included or not. The editors of the *Zeitschrift* had to allow for some disagreement on its pages to cater to the wide-ranging membership of the *DMG*. But the sensibilities of this same diverse group of people made it necessary to prevent extended, heated debates. Zarncke also tried to find a balance between the *Centralblatt*'s primary aim of being a review journal and allowing critical responses. He adopted the same solution as the editors of the *Zeitschrift*: a maximum of one response by the reviewee and one reply by the reviewer.

This widely shared aversion to prolonged debate among the editors of nineteenth-century German scientific journals has also been observed by others. In an analysis of the *Historische Zeitschrift*, Andreas Fahrmeir notes that "address and reply" were remarkably rare.[178] Lynn Nyhart observes that Carl Theodor von Siebold, the editor of the *Zeitschrift für wissenschaftliche Zoologie*, did not tolerate polemic contributions. Such an article "was to be shunned as an effort that did not contribute anything of substance to science and as conduct unbecoming a gentleman and a scientist."[179]

All this suggests that most nineteenth-century German editors did not primarily perceive of their journals as platforms for critical independence. They projected an image of scholarly consensus and harmony rather than one of continuous criticism and debate. This attitude is perhaps best captured in Nyharts summary of Siebold's conception of his *Zeitschrift*, which he considered to be "a public repository of scientific research—fact and theories, but especially facts."[180] Even if editors were not able to simultaneously express their loyalty to their publishers, their readers, and their contributors, their collective attempts to present some sort of "idyllic still life" on the pages of their journals uncovers an attachment to a greater loyalty: a commitment to an ideal of scholarship as a harmonious, collaborative effort in which occasional ungentlemanly disagreements lose their urgency in the light of the ever-growing body of knowledge. This ideal of scholarly community will be further investigated in the next chapter, which looks at the actual content of book reviews.

Notes

1. Mario Biagioli, "From Book Censorship to Academic Peer Review," *Emergences* 12, no. 1 (2002): 14.
2. Dorothy Stimson, *Scientists and Amateurs: A History of the Royal Society* (New York, NY: Henry Schuman, 1948), 141.
3. Charles Bazerman, *Shaping Written Knowledge: The Genre and Activity of the Experimental Article in Science* (Madison, WI: University of Wisconsin Press, 1988), 137.
4. David A. Kronick, "Peer Review in 18th-Century Scientific Journalism," *Journal of the American Medical Association* 263, no. 10 (1990): 1321.
5. John C. Burnham, "The Evolution of Editorial Peer Review," *Journal of the American Medical Association* 263, no. 10 (1990): 1324.
6. Christopher C. Booth, "The Origin and Growth of Medical Journals," *Annals of Internal Medicine* 113, no. 5 (1990): 402.
7. Drummond Rennie, "Editorial peer review: its development and rationale," in *Peer Review in Health Sciences*, ed. Fiona Godlee and Tom Jefferson (London: BMJ Books, 1999), 6.
8. Stephen Lock, *A Difficult Balance: Editorial Peer Review in Medicine* (London: The Nuffield Provincial Hospitals Trust, 1985), 3.
9. Joachim Kirchner, *Das deutsche Zeitschriftenwesen: Teil II: vom Wiener Kongress bis zum Ausgange des 19. Jahrhunderts* (Wiesbaden: Otto Harrasowitz, 1962), passim.
10. William H. Brock, *Justus von Liebig: The Chemical Gatekeeper* (Cambridge: Cambridge University Press, 1997), 52–53 and J.P. Phillips, "Liebig and Kolbe, Critical Editors," *Chymia* 11 (1966): 89–97.
11. Thomas Hapke, *Die Zeitschrift für Physikalische Chemie: Hundert Jahre Wechselwirkung zwischen Fachwissenschaft, Kommunikationsmedium und Gesellschaft* (Herzberg: Traugott Bautz, 1990), 22–47.
12. J.B. Morrell, "The chemist breeders: the research schools of Liebig and Thomas Thomson," *Ambix* 19, no. 1 (1972): 5–6.
13. Lynn K. Nyhart, "Writing Zoologically: The Zeitschrift für wissenschaftliche Zoologie and the Zoological Community in Late Nineteenth-Century Germany," in *The Literary Structure of Scientific Argument: Historical Studies*, ed. Peter Dear (Philadelphia, PA: University of Philadelphia Press, 1991), 44.
14. Leizpig University provides an overview of these dissertations at https://home.uni-leipzig.de/wundtbriefe/wwcd/chronos/chronos.htm.
15. Wilhelm Wundt to Emil Kraepelin, 14 October 1880, in *Der Briefwechsel zwischen Wilhelm Wundt und Emil Kraepelin*, ed. Holger Steinberg (Bern: Hans Huber, 2002), 39–40.

16. Chiara Russo Kraus, "Back to the origins of the repudiation of Wundt: Oswald Külpe and Richard Avenarius," *Journal of the History of the Behavioral Sciences* 53, no. 1 (2017): 30.
17. For example, Richard Avenarius to Wilhelm Wundt, 5 May 1877, Universitätarchiv Leipzig (UAL): NA Wundt/III/1001-1100/1015/53-56 and Richard Avenarius to Wilhelm Wundt, 21 September 1879, UAL: NA Wundt/III/1001-1100/1019/69-72.
18. Richard Avenarius to Wilhelm Wundt, 31 December 1878, UAL: NA Wundt/III/1001-1100/1017/61-64.
19. Richard Avenarius to Wilhelm Wundt, 19 December 1880, UAL: NA Wundt/III/1001-1100/1021/77-80.
20. Richard Avenarius to Wilhelm Wundt, 31 December 1880, UAL: NA Wundt/III/1001-1100/1022/10-14.
21. Wilhelm Wundt to Emil Kraepelin, 14 October 1880, in *Der Briefwechsel zwischen Wilhelm Wundt und Emil Kraepelin*, 40.
22. Wilhelm Wundt to Emil Kraepelin, 4 August 1881, in Steiner, *Der Briefwechsel zwischen Wilhelm Wundt und Emil Kraepelin*, 58.
23. See: https://home.uni-leipzig.de/wundtbriefe/wwcd/chronos/chronos.htm.
24. See also: Cezary W. Domanski, "A biographical note on Max Friedrich (1856–1887), Wundt's first PhD student in experimental psychology," *Journal of the History of the Behavioral Sciences* 40, no. 3 (2004).
25. David A. Kronick, *A History of Scientific and Technical Periodicals: The Origins and Development of the Scientific and Technological Press, 1665–1790* (New York, NY: The Scarecrow Press, 1962), 193–200.
26. Thomas H. Broman, "J.C. Reil and the "Journalization" of Physiology," in *The Literary Structure of Scientific Argument: Historical Studies*, ed. Peter Dear (Philadelphia, PA: University of Pennsylvania Press, 1991), 23–24.
27. Richard Avenarius to Wilhelm Wundt, 19 December 1880, UAL: NA Wundt/III/1001-1100/1021/77-80.
28. Richard Avenarius to Wilhelm Wundt, 31 December 1880, UAL: NA Wundt/III/1001-1100/1022/10-14.
29. Rudolf Engelmann to Wilhelm Wundt, 6 June 1881, UAL: NA Wundt/III/1601-1700/1681-1. Engelmann's emphasis.
30. Ibid. and Rudolf Engelmann to Wilhelm Wundt, 8 November 1882, UAL: NA Wundt/III/1601-1700/1681-2.
31. Wilhelm Wundt to Emil Kraepelin, 17 December 1880, in Steiner, *Der Briefwechsel zwischen Wilhelm Wundt und Emil Kraepelin*, 40.
32. Emanuel Reinecke to Wilhelm Wundt, 7 February 1890, UAL: NA Wundt/III/1601-1700/1681-5.

33. Rudolf Engelmann to Wilhelm Wundt, 6 June 1881, UAL: NA Wundt/III/1601–1700/1681–1
34. Kirchner, *Das Deutsche Zeitschriftenwesen*, 250.
35. Wilhelm Wundt to Oswald Külpe, 25 December 1904, UAL: NA Wundt/III/401–500/405/25–28.
36. I discuss the case of Lange more extensively in: Christiaan Engberts, "Wilhelm Wundt's Critical Loyalty: Balancing Gendered Virtues Among Early Experimental Psychologists," in *Gender, Embodiment, and the History of the Scholarly Persona: Incarnations and Contestation*, ed. Kirsti Niskanen and Michael J. Barany (London: Palgrave Macmillan, 2021), 10–12.
37. M. v. Laue, "Dr. Ludwig Lange. 1863–1936. (Ein zu Unrecht Vergessener)," *Die Naturwissenschaften* 35, no. 7 (1948): 194.
38. Ludwig Lange to Wilhelm Wundt, 9 June 1885, UAL: NA Wundt/III/401–500/433a/171–182.
39. His dissertation consisted of these two papers: Ludwig Lange, "Die geschichtliche Entwicklung des Bewegungsbegriffes und ihr voraussichtliches Endergebnis. I.," *Philosophische Studien* 3, no. 3 (1886): 337–419 and Ludwig Lange, "Die geschichtliche Entwicklung des Bewegungsbegriffes und ihr voraussichtliches Endergebnis. II. (Schluss)," *Philosophische Studien* 3, no, 4 (1886): 643–691.
40. Laue, "Dr. Ludwig Lange," 194–195.
41. F. Kiesow, "F. Kiesow," in *A history of psychology in autobiography*, volume I, ed. Carl Murchison (New York, NY: Russell & Russell, 1961), 172.
42. Ludwig Lange to Wilhelm Wundt, 30 December 1887, UAL: NA Wundt/III/401–500/433c/187–190.
43. For an overview of Lange's contributions to the *Studien*, see: Hans Lindau, *Philosophische Studien herausgegeben von Wilhelm Wundt: Namen- und Sachregister zu Band I-XX* (Leipzig: Wilhelm Engelmann, 1904), 7–8.
44. Ludwig Lange to Wilhelm Wundt, 15 March 1919, UAL: NA Wundt/III/401–500/434a/253–256.
45. Julius Merkel, "Die zeitlichen Verhältnisse der Willenstätigkeit," *Philosophische Studien* 2, no. 1 (1883).
46. Wilhelm Wundt to Julius Merkel, 28 March 1886, UAL: NA Wundt/III/601–700/665/307–312. Lorenz finished his dissertation in 1885, Starke in 1886.
47. Wilhelm Wundt to Julius Merkel, 5 October 1886, UAL: NA Wundt/III/601–700/673/335–338.
48. For an overview of Merkel's contributions to the *Studien*, see: Lindau, *Philosophische Studien*, 9–11.

49. Merkel is mentioned as mathematics teacher in the annual report of a Zittau high school in 1889: Johannes Schütze, *Jahresbericht des Königl. Realgymnasium mit höherer Handelsschule zu Zittau für das Schuljahr 1888/89* (Zittau, 1889), 32. The fact he started writing from Zittau as soon as 1887. Suggests that he already taught there in that year: Julius Merkel to Wilhelm Wundt, 8 October 1887, UAL: NA Wundt/III/601–700/674/339–346. He appears to have been a loyal employee: in 1915 he still worked at the same school: Ernst Korselt, *Jahresbericht des Königlichen Realgymnasiums mit Höherer Handelsschule in Zittau für das Schuljahr Ostern 1914 bis Ostern 1915* (Zittau, 1915), 22.
50. He is not mentioned in the list mentioned in note 14. In an autobiographical essay, however, he mentions that he worked on his dissertation in Wundt's institute and the Wundt was one of its assessors: Kiesow, "F. Kiesow," 177.
51. David Kent Robinson, Wilhelm Wundt and the Establishment of Experimental Psychology (PhD diss., University of California, Berkeley, 1987), 137.
52. For an overview of Kiesow's contributions to the *Studien*, see: Lindau, *Philosophische Studien*, 5–6.
53. Wilhelm Wundt to Friedrich Kiesow, 19 November 1897, UAL: NA Wundt/III/201–300/214/59–62.
54. Wilhelm Wundt to Friedrich Kiesow, 15 February 1902, UAL: NA Wundt/III/201–300/220/89–92. August Kirschmann was another student of Wundt without *Abitur*. In 1902 was appointed as a full professor in Toronto in 1902.
55. Friedrich Kiesow to Wilhelm Wundt, 31 December 1919, UAL: NA Wundt/III/201–300/232/141–148.
56. Wilhelm Wundt to Gottlob Friedrich Lipps, 4 February 1900, UAL: NA Wundt/III/501–600/579/559–562.
57. Wilhelm Wundt to Gottlob Friedrich Lipps, 15 February 1902, UAL: NA Wundt/III/501–600/585/583–586.
58. For an overview of Lipps' contributions to the *Studien*, see: Lindau, *Philosophische Studien*, 8–9.
59. Wilhelm Wundt to Gottlob Friedrich Lipps, 15 February 1894, UAL: NA Wundt/III/501–600/569/515–518 and Wilhelm Wundt to Gottlob Friedrich Lipps, 1 December 1898, UAL: NA Wundt/III/501–600/576/547–550.
60. Richard Pischel, Franz Praetorius, Ludolf Krehl and Ernst Windisch, *Die Deutsche Morgenländische Gesellschaf 1845–1895. Ein Ueberblick* (Leipzig: F.A. Brockhaus, 1895), 41.
61. Sabine Mangold, *Eine "weltbürgerliche"Wissenschaft—Die deutsche Orientalistik im 19. Jahrhundert* (Stuttgart: Franz Steiner, 2004), 66–67.

62. An incomplete list of Fleischer's most famous students is provided in: Johann Fück, *Die arabischen Studien in Europa bis in den Anfang des 20. Jahrhunderts* (Leipzig: Otto Harrassowitz, 1955), 172.
63. Holger Preissler, "Die Anfänge der Deutschen Morgenländischen Gesellschaft," *Zeitschrift der Deutschen Morgenländischen Gesellschaft* 145, no. 2 (1995): 247.
64. Theodor Nöldeke to Michael Jan de Goeje, 25 September 1858, UBL: BPL 2389.
65. Michael Jan de Goeje to Theodor Nöldeke, 21 October 1864, UBL: BPL 2389.
66. Elizabeth Yale, *Sociable Knowledge: Natural History and the Nation in Early Modern Britain* (Philadelphia, PA: University of Pennsylvania Press, 2016), 65.
67. Michael Jan de Goeje to Theodor Nöldeke, 9 November 1862, UBL: BPL 2389.
68. Heinrich Leberecht Fleischer to Michael Jan de Goeje, 17 March 1863, UBL: BPL 2389.
69. Heinrich Leberecht Fleischer to Theodor Nöldeke, 29 April 1879, Universitätsbibliothek Tübingen (hereafter UBT): Md 782 A 68.
70. Heinrich Leberecht Fleischer to Theodor Nöldeke, 4 August 1879, UBT: Md 782 A 68.
71. Theodor Nöldeke, "Their und Mensch vor dem König der Genien," *Literarische Centralblatt* (1879): 1259–1260.
72. Patrick M. Graham, "The Discovery and Reconstruction of the Mesha Inscription," in *Studies in the Mesha inscription and Moab*, ed. Andrew Dearman (Atlanta, GA: Scholars Press, 1989), 42.
73. Martin Heide, "The Moabitica and Their Aftermath: How to Handle a Forgery Affair with an International Impact," in *New inscriptions and seals relating to the biblical world*, ed. Meir Lubetski and Edith Lubetski (Atlanta, GA: Society of Biblical Literature, 2012), 194–195.
74. Konstantin Schlottmann, "Neue Moabitische Funde und Räthsel: Erster Bericht," *Zeitschrift der Deutschen Morgenländischen Gesellschaft* 26 (1872): 393–416.
75. "Protokollarischer Bericht über die in Halle a/S. am 28. Sept. 1872 abgehaltene Generalversammlung d. D.M.G.," *Zeitschrift der Deutschen Morgenländischen Gesellschaft* 27 (1873): vi.
76. Heide, "The Moabitica and Their Aftermath," 198.
77. Albert Socin, "Über Inschriftfälschungen," *Zeitschrift der Deutschen Morgenländischen Gesellschaft* 27 (1873): 133–135.
78. Heinrich Leberecht Fleischer to Theodor Nöldeke, 5 January 1875, UBT: Md 782 A 68.

79. Heinrich Leberecht Fleischer to Albert Socin, 22 October 1873, UBT: Md 782 A 68. This letter can be found among the Fleischer's letter to Nöldeke. It is most likely a copy of the letter he sent to Socin.
80. Heinrich Leberecht Fleischer to Theodor Nöldeke, 5 January 1875, UBT: Md 782 A 68.
81. Heide, "The Moabitica and Their Aftermath," 200.
82. Konstantin Schlottmann, "Ueber die Aechtheit der Moabitischen Alterthümer," *Zeitschrift der Deutschen Morgenländischen Gesellschaft* 28 (1874): 171–184.
83. For example: Ludwig Diestel, "Die moabitischen Alterthümer," *Jahrbücher für deutsche Theologie* 21 (1876): 451–473 and Adolf Koch, *Moabitisch oder Selimisch? Die Frage der moabitischen Altertümer* (Stuttgart, Schweizerbart, 1876).
84. Emil Kautzsch and Albert Socin, *Die Aechtheit der moabitischen Altertümer geprüft* (Strasbourg: Trübner, 1876).
85. Theodor Nöldeke, "Die maobitschen Fälschungen," *Deutsche Rundschau* 6 (1876): 447–451; Theodor Nöldeke, "Koch Ad, Moabitisch oder Selimisch?," *Literarische Centralblatt* 13 (1876): 425–428; Theodor Nöldeke, "'Bädeker' im Morgenlande," *Im neuen Reich. Wochenschrift für das Leben des deutschen Volkes in Staat, Wissenschaft und Kunst* 5, no. 2 (1875): 15–20.
86. Heinrich Leberecht Fleischer to Theodor Nöldeke, 21 July 1875, UBT: Md 782 A 68.
87. Michael Jan de Goeje to Theodor Nöldeke, 19 November 1876, 7 January 1878 and 4 December 1882, UBL: BPL 2389 and Theodor Nöldeke to Michael Jan de GOeje, 11 October 1876, UBL: Or. 5585e.
88. August Fischer of the executive board of the *Deutsche Morgenländische Gesellschaft*, 9 November 1905, Universitätsarchiv Halle-Wittenberg (hereafter UAHW): Rep. 90:67 and August Fischer to the executive board of the *Deutsche Morgenländische Gesellschaft*, 4 April 1908, UAHW: Rep. 90:70.
89. Mangold, *Eine "weltbürgerliche" Wissenschaft*, 94–95.
90. Gustav Jahn to the executive board of the *Deutsche Morgenländische Gesellschaft*, 8 November 1905, UAHW: Rep. 90:67.
91. Franz Praetorius to the executive board of the *Deutsche Morgenländische Gesellschaft*, 9 November 1905, UAHW: Rep. 90:67.
92. August Fischer to the executive board of the *Deutsche Morgenländische Gesellschaft*, 10 November 1905, UAHW: Rep. 90:67.
93. August Fischer to the executive board of the *Deutsche Morgenländische Gesellschaft*, 12 November 1905, UAHW: Rep. 90:67.
94. Ernst Windisch to the executive board of the *Deutsche Morgenländische Gesellschaft*, 12 November 1905, UAHW: Rep. 90:67.

95. August Fischer to the executive board of the *Deutsche Morgenländische Gesellschaft*, 9 November 1905, UAHW: Rep. 90:67.
96. August Fischer, "Redakteurglossen," *Zeitschrift der Deutschen Morgenländischen Gesellschaft* 59 (1905): 442.
97. Theodor Nöldeke to Michael Jan de Goeje, 8 July 1905, UBL: BPL 2389 and Theodor Nöldeke to Ignaz Goldziher, 10 July 1905, Library and Information Centre of the Hungarian Academy of Sciences (hereafter MTAK): GIL/32/01.
98. Michael Jan de Goeje to Theodor Nöldeke, 17 July 1905, UBL: BPL 2389.
99. August Fischer to Michael Jan de Goeje, 19 July 1905, UBL: BPL 2389.
100. August Fischer to Michael Jan de Goeje, 18 September 1905, UBL: BPL 2389.
101. "Protokollarischer Bericht über die am 2. Und 3. Oktober 1905 zu Hamburg abgehaltene allgemeine Versammlung der D.M.G.," *Zeitschrift der Deutschen Morgenländischen Gesellschaft* 59 (1905): lxxii.
102. Jakob Barth, *Sprachwissenschaftliche Untersuchungen zum Semitischen*, (Leipzig: J.C. Hinrichs'sche Buchhandlung, 1907), 30.
103. August Fischer, "Allerlei von J. Barth 'verbesserte' arabische Dichterstellen," *Zeitschrift der Deutschen Morgenländischen Gesellschaft* 61 (1907): 938.
104. August Fischer to the full board of the Deutsche Morgenländische Gesellschaft, 4 April 1908, UAHW: Rep. 90:70.
105. Franz Praetorius to the full board of the Deutsche Morgenländische Gesellschaft, no date [April 1908], UAHW: Rep. 90:70.
106. Theodor Nöldeke to the full board of the Deutsche Morgenländische Gesellschaft, 10 April 1908, UAHW: Rep. 90:70.
107. Theodor Nöldeke to Michael Jan de Goeje, 11 April 1908, UBL: BPL 2389.
108. August Fischer, "Erklärung," *Zeitschrift der Deutschen Morgenländischen Gesellschaft* 62 (1908): 203.
109. Theodor Nöldeke to Michael Jan de Goeje, 19 May 1908, UBL: BPL 2389.
110. "Personalnachrichten," *Zeitschrift der Deutschen Morgenländischen Gesellschaft* 62 (1908): xl.
111. "Protokollarischer Bericht über die Mittwoch den 14. Okt. 1908 zu Leipzig abgehaltene Allgemeine Versammlung der D.M.G.," *Zeitschrift der Deutschen Morgenländischen Gesellschaft* 62 (1908) xlvi-xlvii.
112. August Fischer to the full board of the DMG, 4 April 1908, UAHW: Rep. 90:70.
113. "Redaktionswechsel," *Zeitschrift der Deutschen Morgenländischen Gesellschaft* 64 (1910): xxxvii.

114. "Mitgliedernachrichten," *Zeitschrift der Deutschen Morgenländischen Gesellschaft* 64 (1910): xlv; Theodor Nöldeke, "Zum 'Buch der Gesetze der Länder'," *Zeitschrift der Deutschen Morgenländischen Gesellschaft* 64 (1910): 555–560.
115. Kirchner, *Das deutsche Zeitschriftenwesen*, 70–71, 242.
116. Thomas Lick, *Friedrich Zarncke und das 'Literarische Centralblatt für Deutschland.' Eine buchgeschichtliche Untersuchung* (Wiesbaden: Harrassowitz, 1993), 13–14.
117. Stefan Rebenich, *Theodor Mommsen: Eine Biographie* (Munich: C.H. Beck, 2002) 70–71.
118. Lick, *Friedrich Zarncke*, 48–51.
119. Ludwig Avenarius, *Avenarianische Chronik: Blätter aus drei Jahrhunderten einer deutschen Bürgerfamilie* (Leipzig: O.R. Reisland, 1912), 85, 89.
120. Avenarius, *Avenarianische Chronik*, 86.
121. Eduard Avenarius to Friedrich Zarncke, 20 March 1862, Universitätsbibliothek Leipzig (hereafter UBLE): NL 249/1/A/570.
122. Eduard Avenarius to Friedrich Zarncke, 15 May 1862, UBLE: NL 249/1/A/569.
123. Eduard Avenarius to Friedrich Zarncke, 28 May 1864, UBLE: NL 249/1/A/565.
124. Ibid.
125. Waldemar Stössel, "Anton Klette," in *Geschichte der Universitätsbibliothek Jena* (Weimar: Hermann Böhlhaus Nachfolger, 1958), 490–492.
126. Eduard Avenarius to Friedrich Zarncke, 22 March 1874, UBLE: NL/249/1/A/559.
127. Ibid.
128. Eduard Avenarius to Friedrich Zarncke, 11 February 1876, UBLE: NL/249/1/A/537.
129. Eduard Avenarius to Friedrich Zarncke, 1 March 1875, UBLE: NL/249/1/A/535.
130. Wilhelm Wundt to Friedrich Zarncke, 3 January 1876, UAL: NA Wundt/III/966.
131. Eleonore Wundt, *Wilhelm Wundts Werk: Ein Verzeichnis seiner sämtlichen Schriften* (Munich: C.H. Beck, 1927), 12–15.
132. Theodor Nöldeke to Michael Jan de Goeje, 30 November 1867, UBL: BPL 2389.
133. James A. Secord, *Victorian Sensation: The Extraordinary Publication, Reception, and Secret Authorship of Vestiges of the Natural History of Creation* (Chicago, IL: University of Chicago Press, 2000), 367.
134. Friedrich Zarncke, "An unsere Leser," *Literarische Centralblatt* 52 (1874), 1722–1723.

135. Mary Ruth Hiller, "The Identification of Authors: The Great Victorian Enigma," in *Victorian Periodicals: A Guide to Research*, ed. Don J. Vann and Rosemary T. VanArsdel (New York (NY): The Modern Language Association of America, 1978), 126.
136. Eduard Avenarius to Friedrich Zarncke, 24 September 1874, UBLE: NL 249/1/A/556.
137. Heinrich Ewald to Friedrich Zarncke, 3 September 1851, UBLE: NL 249/1/E/1051.
138. Eduard Avenarius to Friedrich Zarncke, 1 October 1874, UBLE: NL 249/1/A/555.
139. Eduard Avenarius to Friedrich Zarncke, 10 December 1874, UBLE: NL 249/1/A/553.
140. Ibid. Zarncke used the words 'Lohnschreiber untersten Ranges' and 'Lohnschreiber.'
141. Ibid. Zarncke used the words 'Rechtschaffenes' and 'Gediegenes.'
142. Eduard Avenarius to Friedrich Zarncke, 18 December 1874, UBLE: NL 249/A/545.
143. Stössel, "Anton Klette," 514.
144. Veit & Comp., "Zur Nachricht," *Jenaer Literaturzeitung* 39 (1879): 536–537.
145. Ernst Mehl, "Deutsche Bibliotheksgeschichte," in *Deutsche Philologie im Aufriss*, Band I, ed. Wolfgang Stammler (Berlin/Bielefeld: Erich Schmitt, 1952), 358; Jürgen Storost, "Zu einigen germanistischen Zeitschriften in der zweiten Hälfte des 19. Jahrhunderts," *Zeitschrift für Germanistik*, Neue Folge II (1992): 351.
146. Eduard Avenarius to Friedrich Zarncke, 24 September 1880, UBLE: NL 249/1/A/543.
147. Eduard Avenarius to Friedrich Zarncke, 10 March 1862, UBLE: NL 249/1/A/571.
148. F.A. Brockhaus to Friedrich Zarncke, 22 August 1868, UBLE: NL 249/1/B/2070; D.lbr.ck [Berthold Delbrück], "Kellner, Dr. Camillo, 'kurze Elmentargrammatik der Sanskrit-Sprache'," *Literarische Centralblatt* 30 (1868): 814–815; Camillo Kellner, "Entgegnung," *Literarische Centralblatt* 42 (1868): 1147–1148.
149. F.A. Brockhaus to Friedrich Zarncke, 6 November 1871, ULBE: NL 249/1/B/2077; F., "Hergt, Dr. C.E., Geographie des Gelobten Landes'," *Literarische Centralblatt* 36 (1871): 907–908.
150. F.A. Brockhaus to Friedrich Zarncke, 9 December 1871, UBLE: NL 249/1/B/2078.
151. For example: Franz Delitzsch to Friedrich Zarncke, 1 February 1872, UBLE: NL249/1/D/279; FD [Franz Delitzsch] "Ebers, Dr. Georg,

Prof., Durch Gosen zum Sinai," *Literarische Centralblatt* 9 (1873): 257–260.
152. F.A. Brockhaus to Friedrich Zarncke, 2 March 1869, UBLE: NL 249/1/B/2072.
153. F.A. Brockhaus to Friedrich Zarncke, 17 March 1869, UBLE: NL 249/1/B/2075.
154. F.A. Brockhaus to Friedrich Zarncke, 30 November 1872, UBLE: NL 249/1/B/2081; F.A. Brockhaus to Friedrich Zarncke, 8 April 1880, UBLE: NL 249/1/B/2082.
155. Die Redaction des Literarischen Centralblattes [Friedrich Zarncke], "Literarische Anzeigen," *Literarische Centralblatt* 5 (1853): 91–92.
156. Ibid.
157. See page 77–80 of this chapter.
158. Lick, *Friedrich Zarncke*, 313.
159. Bernhard Maier, *Gründerzeit der Orientalistik: Theodor Nöldekes Leben und Werk im Spiegel seiner Briefe* (Stuttgart: Franz Steiner, 2004), 430–438; Eleonore Wundt, *Wilhelm Wundts Werk: Ein Verzeichniss seiner sämtlichen Schriften* (Munich: C.H. Beck, 1927), 7–16.
160. Theodor Nöldeke to Michael Jan de Goeje, 29 October 1865, UBL: BPL 2389.
161. Alfred Gutschmid to Friedrich Zarncke, 4 April 1866, UBLE: NL 249/1/G/1379.
162. Franz Delitzsch to Friedrich Zarncke, 16 June 1872, UBLE: NL 249/1/D/280.
163. Franz Delitzsch to Friedrich Zarncke, 16 February 1874, UBLE NL 249/1/D/282.
164. Franz Delitzsch to Friedrich Zarncke, 21 June 1882, UBLE: NL 249/1/D/297; Th. N. [Theodor Nöldeke], "Libri Danielis Ezrae et Nehemine," *Literarische Centralblatt* 34 (1882): 1137–1140.
165. Wilhelm Wundt to Friedrich Zarncke, 26 June 1872, UAL: NA Wundt/ III/956.
166. Eduard Avenarius to Friedrich Zarncke, 22 March 1874, UBLE: NL 249/1/A/559.
167. Richard Avenarius to Friedrich Zarncke, 23 May 1872, UBLE: NL 249/1/A/606; [Richard Avenarius], "Steinthal, Dr. H., Prof., Einleitung in die Psychologie der Sprachwissenschaft," *Literarische Centralblatt* 23 (1872): 599–603.
168. Georg Hoffmann to Friedrich Zarncke, 15 February 1882, UBLE NL 249/1/H/2161; G.H. [Georg Hoffmann], "Nöldeke, Theod., kurzgefasste syrische Grammatik," *Literarische Centralblatt* 10 (1882), 318–322.

169. Suzanne L. Marchand, *German Orientalism in the Age of Empire: Religion, Race, and Scholarship* (Cambridge: Cambridge University Press, 2009), 148.
170. *Ignaz Goldziher: Tagebuch*, ed. Alexander Schreiber (Leiden: Brill, 1978), 26.
171. Alfred von Gutschmid to Friedrich Zarncke, 23 April 1873, UBLE: NL 249/1/G/1413.
172. A v. G. [Alfred von Gutschmid], "Vámbéry, Herm., Geschicihte BOchara's oder Transoxanien," *Literarische Centralblatt* 19 (1873): 577–590.
173. Hermann Vámbéry and Alfred von Gutschmid, "Erwiderungen," *Literatische Centralblatt* 24 (1873)L 763–768.
174. Ibid.
175. Alfred von Gutschmid to Friedrich Zarncke, 2 June 1873, UBLE: NL 249/1/G/1414.
176. Based on the list of members in the 1875 issue of the *Zeitschrift*.
177. Eduard Avenarius to Friedrich Zarncke, 10 March 1862, UBLE: NL/249/1/A/571.
178. Andreas Fahrmeir, "Ort des Konseses oder Historische Streitschrift? Zur Geschichte des 19. und 20. Jahrhunderts in der Historischen Zeitschrift," *Historische Zeitschrift* 289 (2009): 201.
179. Nyhart, "Writing Zoologically," 54.
180. Ibid., 45.

CHAPTER 4

A Review of Reviews: Criticism and Community Formation in Book Reviews

The Genre of the Book Review

"You worthy reviewers place yourselves on […] a pedestal and speak too much like chief justices about poor clients who stand trial before you. […] You have too much of an eye for the defects and do not place enough value on the good. […] That makes a writer or publisher bitter, you see."[1] Jan de Goeje's protests to a highly critical review of a book by the Vienna orientalist Alfred von Kremer by Theodor Nöldeke illustrate his unease with the genre. In his eyes, reviewers tended to be needlessly critical of their peers' imperfect but ultimately praiseworthy efforts. Nöldeke, however, was not impressed by his friend's accusations. He argued that Kremer's book deserved all of his harsh criticism and stated that on this issue he felt "more fully in the right than ever."[2] Still, the editorial considerations described in the previous chapter suggest that De Goeje's preference for more sympathetic reviews was shared by many of his contemporaries.

The genre of the—largely sympathetic—book review had developed as the by-product of the growing importance of book publishing. An increasing appreciation of innovation rather than mere erudition encouraged scholars across early modern Europe to publish their novel findings as soon as possible.[3] This publishing boom was further accelerated by technological developments in the book printing industry that caused the

© The Author(s), under exclusive license to Springer Nature Switzerland AG 2022

C. Engberts, *Scholarly Virtues in Nineteenth-Century Sciences and Humanities*, https://doi.org/10.1007/978-3-030-84566-7_4

European book production to triple between 1700 and 1770.[4] During these years, book reviews, bibliographies, and book fair catalogues provided an overview of the flood of newly published scientific literature. Early modern scholarly periodicals were the first to print book reviews and bibliographies. From the second half of the eighteenth century onwards, specialized review journals would become an increasingly prominent feature of scholarly life.[5]

Early modern book reviews were decisively shaped by the fact that their primary function was not to assess their scholarly merit but rather to give an overview of newly published works. They usually simply summarized the contents of the work under review without judging its worth.[6] By the end of the eighteenth century, however, new reviewing styles emerged. In 1799, the theologian Johann Christoph Greiling recognized three ways of reviewing: announcements, reviews in a narrow sense, and reviews in a broad sense. "Reviewing can be understood in a historical and a philosophical meaning. In the first case, it would mean: to state the contents of a book: reporting. In the philosophical meaning, however, reviewing must mean: to examine the spirit of a book on the basis of the principles of the discipline to which it […] belongs. The first type of review is called announcements, the second reviews in the narrow sense. Reviewing in its wider meaning brings together both types."[7] The review in a narrow sense, but especially the review in a wider sense, encouraged the reviewer to provide a critical assessment of the reviewed work.

Greiling's distinctions draws attention to the way in which early book reviews differ from those published today. In a recent study, Ken Hyland notes that all contemporary book reviews are predominantly what Greiling would have called "reviews in the narrow sense." In his eyes, modern-day book reviews constitute "essentially an evaluative genre where writers judge a text on its academic quality, clarity, integrity and value to the field."[8] Rosa Lorés-Sanz agrees with this characterization and even states that the genre as we know it today is evaluative "by definition."[9] The book review as a straightforward announcement, however, seems to have become something of the past.

Another type of review, that was quite common in the nineteenth century as well, has also become somewhat old-fashioned: the book review as an original contribution to scholarship. A lot of nineteenth-century reviewers refused to limit themselves to simply summarizing or assessing the merits of a book. They used their reviews to share their own insights too. The French sociologist Émile Durkheim, for instance, was widely known to use his book reviews "as a platform for the elucidation of his own theories and for rebuttal of the attacks of his critics."[10] Late nineteenth-century German historians, likewise, used book reviews to present their "own points of view, concepts and current research."[11] By the end of the nineteenth century, the book review had developed into a very diverse genre that could contain different measures of summarizing, evaluation, and the presentation of the reviewer's own insights.

In this chapter, I will take a closer look at the book reviews that Theodor Nöldeke and Wilhelm Wundt published in the *Literarisches Centralblatt* between 1871 and 1880. The editorials of Friedrich Zarncke show that his conception of what type of reviews suited the journal changed over the years. In the first issue of the *Centralblatt*, he advocated the early modern ideal of the review as a summary: "The journal [...] has given itself the task to provide a complete [...] overview of the full literary activity in Germany [...] and it will provide explanatory notes and short reports of all important books."[12] Twenty-five years later, however, Zarncke emphasized the importance of critical evaluation. He argued that even the summarizing reviews in his journal were aimed at "a wider and higher purpose: to carry the sense for correct and exact methods of research into the widest circles."[13] A quick glance at the pages of the *Centralblatt* suffices to draw the conclusion that it indeed presented a mix of announcements, reviews in the narrow and wider sense, and even a few original contributions masquerading as book reviews.

However, even the largely descriptive reviews contain evaluative content: studies are characterized as, for example, interesting, relevant, or thorough, and authors are praised for their work ethics or careful attention to detail. The evaluative strains that run through almost all reviews by Nöldeke and Wundt provide a promising starting point to answering a myriad of questions. What were the most common reasons to criticize a book? What were the most frequently occurring criticisms of the books' authors? What qualities of books and authors were singled out for praise? And, last but not least, what does this mixture of criticism and praise suggest about the ways in which book reviews were used to express critical independence and collegial loyalty?

In the first half of this chapter, I will investigate whether Nöldeke and Wundt heaped the same amount of criticism and praise on different types of authors. I will look into authors working in different disciplines, of different nationalities, with different religious backgrounds, and with different academic backgrounds. I initially planned to add gender as another analytical category, but of the more than 200 reviews of Nöldeke and Wundt, only one deals with the work of a female author: the dissertation of the Austrian psychologist Susanna Rubinstein.[14] The second half of the chapter changes the focus from the reviewers' attitudes towards different types of authors to the actual criticism and praise in their reviews. I will pay attention to both the qualities that Nöldeke and Wundt valued in the reviewed works and the scholarly virtues and vices that they attributed to the authors of the books.

In short, the first half of the chapter will set out to whom the reviewers were loyal and of whom they were critical. The second half will discuss what this teaches us about the virtues of collegial loyalty and critical independence. These observations will then be the starting point of a linguistic analyses of politeness that further illustrates the way in which book reviews contributed to the creation and maintenance of a scholarly community.[15]

Assessing Different Disciplines

The *Centralblatt* organized its reviews into seventeen thematic sections that included all major disciplines taught at German universities and a section for miscellaneous reviews.[16] The expertise of Nöldeke and Wundt allowed them to contribute to different sections. Because of the traditional close ties between Old Testament studies and Semitic languages, Nöldeke was able to comment on both topics. Half of his ninety-six reviews of the 1870s were printed in the *Theology* section and forty were published in the *Linguistics* section. Most of the few remaining reviews were of books on the history of the Islamic world and were featured in the *History* section. Wundt had started contributing to the *Centralblatt* before he had switched from physiology to philosophy. After accepting the Zürich Chair of Inductive Philosophy in 1874, however, he reviewed both medical publications and philosophical works. In the end, seventy-three of his reviews in the 1870s were printed in the *Medicine* section, thirty-two in the *Philosophy* section, and fifteen in the *Natural Sciences* section. In accordance with Zarncke's wishes, most of these reviews were quite sympathetic. Of Nöldeke's 96 reviews, only 12 were unambiguously unfavourable, while 19 of Wundt's 123 reviews were undeniably negative.[17]

Nöldeke and Wundt were not equally critical about books reviewed in each section. Nöldeke was clearly more likely to write an unfavourable review of theological or historical studies than of linguistic works. Three of his critical reviews were of historical studies and no less than eight of his twelve negative reviews were printed in the *Theology* section! Only one of his critical reviews was published in the *Linguistics* section. Because he wrote only seven reviews of historical works, it is not possible to draw any definite and generalizable conclusions about his assessments of historical works. The differences between his theological and linguistic reviews, however, is remarkable. While eight of his forty-eight theological reviews were negative, only one of his forty linguistic reviews was unfavourable. This one critical linguistic review was of a booklet by the Italian attorney Giuseppe Barzilai. While Nöldeke admitted that "each dilettante has the full liberty to play around with scholarly matters and to create a building without a steady fundament with some effort of phantasy and ingenuity," he argued that such works should not be published: "[T]he reader will agree with our advice that Mr. Barzilai should from now on deploy his 'nourishment—consolation—energy'—more purposefully than on [...] printing [...] such a work."[18]

Rudi Paret's characterization of Nöldeke as a strict positivist provides a clue as to why he was so much more critical of theological treatises than of linguistic studies.[19] Nöldeke indeed often criticizes theologians for relying on scriptural authority and church dogma rather than on independent thinking and modern methods of critical analysis. In one review, he complains that "the times when a catholic clergyman could [...] without apostatising his church, examine the Bible with true criticism have long passed. [The author] invariably endorses the official views held by the church."[20] In another review in the same issue of the *Centralblatt*, he mockingly paraphrases a section from a book on the authenticity of the Pentateuch, in which the author piously stated that he knew of "no other authority than that of the church, which leaves every examination aiming for truth the freest manoeuvring room." He then sneers that it would require "much less subjection of reason to believe in the 'authenticity' of the Pentateuch [...] than to believe in the infallibility of the pope."[21] Half a year later, he puts another author in his place for seemingly believing that "scholarly criticism no longer has any value when it contradicts his religious views."[22] Even secular scholars were scorned by Nöldeke when they did not sufficiently distance themselves from the points of view of the church. One author, who is "free from religious-dogmatic prejudice" in Nöldeke's eyes, is still criticized for "clinging too heavily to the accepted tradition."[23]

Wundt's reviews show a similar distinction in the assessment of works from different disciplines. Of his seventy-three reviews in the *Medicine* section, only three were unambiguously unfavourable. Three of Wundt's fifteen contributions to the Natural Sciences section were highly critical as well. These were mostly reviews of books about one of his primary fields of interest, the physiology of perception. A book on Weber's law, for example, was brushed aside as a collection of "striking examples of what [physiologists] should *not* do, when they aim to engage in research."[24] He reserved his harshest criticism for philosophers: of his thirty-two reviews in the *Philosophy* section, no less than thirteen—more than one-third—were largely dismissive.

These numbers do not even do full justice to Wundt's critical stance towards philosophical works because his three unfavourable reviews in the *Medicine* section all deal with topics that would have fit in the *Philosophy* section as well. His main criticism of Ludwig Büchner's *Physiologische Bilder* is that his elaborations about the nature of the soul and consciousness amount to an incompetent philosophical analysis rather than to a solid physiological study: "The author makes it easy for himself with the problem of consciousness. Consciousness has to lie dormant in matter, we don't have the right to ask how and why."[25] Another book reviewed in the *Medicine* section is criticized for its reliance on vitalistic theories about "life energy."[26] In the harshest of his negative reviews in this section, Wundt wipes the floor with a remarkably pious author who argued that "the biblical story about the descent of all people from one couple, should not simply be discarded to the realm of fairy tales."[27] Wundt wryly notes that this author lacks the necessary familiarity with Darwin's theory of evolution.

Wundt's critical reviews in the *Medicine* section give a rather representative picture of his critical comments in the *Philosophy* section. Most of the books he dismisses are criticized for drawing on worldviews that he considers to be speculative and obsolete. He is particularly critical of authors working in the tradition of *Naturphilosophie*. Johannes Volkelt's conception of dreams as "the miraculous and the mystical, the opposite of the laws of the awake consciousness" is brushed aside as the "idolisation of dreams practiced in earlier *naturphilosophische* mysticism."[28] Other authors writing in the same philosophical tradition are dismissed in a similar fashion.[29]

The adherents to other, supposedly obsolete, philosophical traditions are criticized by Wundt as well. One study is harshly judged for the way in which it juxtaposes the assumption that "air, water, certain chemical primordial matter, and heat" are the "external elementary conditions of life" with the classical theory of the four elements.[30] Other books are criticized for drawing on Fichtean or Schopenhauerian idealism.[31] The most avid followers of Hegel were not only in for critical scrutiny but also for mockery: "In this volume, we basically have just an account of the Hegelian logic, that distinguishes itself from the master's dry tone only a little bit by the fact that it has been spiced up abundantly with more or less suitable poetic quotes." After citing some of these quotes, Wundt concludes that these "examples should suffice to show how the author has not failed to bestrew the thorny road of dialectics with many flowers."[32]

In the light of his disagreements with Gustav Theodor Fechner about the merit of Henry Slade's accomplishments, it is not surprising that Wundt was also highly critical of any publications advocating spiritism.[33] He wrote his earliest review of a spiritist book and journal a full year before his correspondence on this topic with Fechner. At this time, he was still convinced that these foreign, far-fetched, theories would not catch on: "In Germany, we can only find two scholarly so-called authorities that are known to be held in high regard in spiritist circles, namely Prof. of Zoology Max Perty in Bern and Prof. of Philosophy Franz Hofmann in Würzburg, and even these men have affiliated themselves with the spiritists' efforts with some reservations."[34]

Occasionally, Wundt also praised an author who drew on a philosophical tradition that he himself rejected. One Hegelian was favourably reviewed for his attempts to bridge the gap between Hegelian speculation and modern scientific psychology: "Even if one cannot agree with the author about all of his views, nobody will put down the lucidly and appealingly written little book without feeling very inspired."[35] Even though Wundt was not an advocate of the insights of Johann Friedrich Herbart, he praised a presentation of his ideas by a loyal disciple because it provided "an understandable exposition which is as faithful as possible and which is also suitable for a wider audience."[36] He even wrote one favourable review of an author drawing on the *naturphilosophische* tradition. Heinrich Baumgärtner's *Die Weltzellen* is recommended to anyone interested in the similarities and differences between Darwin's theory of evolution and *naturphilosophische* conceptions of growth and change.[37]

The above case studies suggest that Nöldeke and Wundt recognized a different degree of agreement about ideals of good scholarship in different disciplines. Within linguistics and medicine, there was a quite clear consensus about these ideals. This consensus will be further investigated in this chapter's sections on reasons for praise and criticism in book reviews.[38]

In the more abstract and speculative fields of theology and philosophy, however, there was no such consensus. Nöldeke and Wundt both defended an ideal of scholarship informed by positivism and empiricism against what they considered to be the obsolete tunnel visions of religious and metaphysical speculation. In Nöldeke's reviews, this appears as the repeated criticism of dogmatic, traditionalist approaches to Old Testament studies. In Wundt's reviews, this was articulated by unfavourable reviews of works drawing on the dubious assumptions of such philosophies as *Naturphilosophie*, vitalism, spiritism, and different strands of idealism.

Nöldeke on Religion and Nationality

The *Centralblatt* published reviews of both German and foreign books. This provides an opportunity to investigate whether German and foreign contributions to scholarship were received differently. The late nineteenth century was, after all, an era of fervent nationalism. In German academic circles, many shared the view that German scholarship was intimately connected to a specific German way of thinking that gave it a clear edge over foreign scholarship.[39] The question, then, is whether this national confidence also translated into more favourable reviews of German scholarship and less favourable reviews of foreign works. A related question is whether there were differences in the assessment of scholars with a different religious background. German research universities had grown, in the words of William Clark, "from the faculties of German Protestant professors."[40] Therefore, it is interesting to compare the reception of works of Protestant and non-Protestant authors as well. This group includes not only Catholics but also Jews. After all, "the place of Jews and Judaism in society and theology was a perennial question."[41]

The position of Jews was especially relevant in one of Nöldeke's main fields of expertise: Old Testament studies. In the light of his aversion of religious—and therefore, in his eyes, uncritical—approaches to this field, the question arises whether he was more critical of Jewish authors than of others. If we accept Robert Irwin's description of Nöldeke as a "racial bigot," at least some traces of antisemitism should be expected.[42] Distrust of Jewish scholarship could further be encouraged by the ideology of *Kulturprotestantismus*, which was widely shared among nineteenth-century German orientalists. Its adherents typically advocated secular teaching and research methods and tended to dismiss their religiously inspired peers as blatantly unscientific.[43] A tacit—or even open—dislike of

Jewish scholars was quite common in all sectors of nineteenth-century, German, academic life.[44]

Jewish scholars were well-represented in nineteenth-century German oriental and Old Testament studies, however. To some extent, they were seen as Orientals themselves, with customs and worldviews that resembled those of biblical peoples. This would make them uniquely suited to mediate between contemporary Europe and the biblical past.[45] Of course, their Hebrew was also expected to be better than that of their non-Jewish peers.[46] At least one modern-day author argues that antisemitic comments were so rare in late nineteenth-century ancient Oriental studies that "no evidence can be found for the exclusion of Jews in this field."[47] In the light of the absence, or at least extreme rarity, of Jewish full professors in Oriental or Old Testament studies at European universities, this assessment is probably a bit too optimistic.[48] However, Nöldeke's reflections on his many Jewish students suggest that he was not very prejudiced against them: "I now have two more Jews in my audience [...]. It seems as if through time my lectures change into an actual seminar for higher Judaism. Well, if people are this industrious, it is fine with me!"[49]

Some of Nöldeke's reviews, however, were quite critical of Jews and Judaism. This is particularly notable in his discussions of books about Judaism in the nineteenth century. His review of the first volume of Abraham Geiger's posthumously published essays is an odd mix of praise and criticism. He calls him an "educated, brilliant, erudite, humane, yet spirited man" and compliments the way in which he does justice to "that which is unchangeable and holy to the rigid old believers" while acknowledging that Judaism has not always developed in "a progressive way."[50] In the same breath, however, he accuses Geiger of incorrectly retracing too many features of contemporary life to Jewish traditions. What is more, he also attacks the Jewish religion as incompatible with modern life in three respects: its dietary laws, the "unreasonable strictness" of the Sabbath, and the practice of circumcision. In Nöldeke's eyes, Geiger's attachment to Judaism was at odds with his manifold praiseworthy qualities.

A similar ambivalence can be found in Nöldeke's review of Seligmann Meyer's objections to a series of antisemitic articles by Hermann Messner.[51] He clearly sided with Meyer. In his eyes, Messner's articles contained "a judgement of modern Judaism that is as loveless as it is ignorant." Therefore, it was easy for Meyer to counter these "superficial and hateful allegations." At the same time, however, he argues that Meyer should be more critical of his own faith and accuses him of "an apologetic style, that tends to praise everything Jewish as such" and a general unwillingness to "acknowledge that the tragic histories of the Jews are largely based on their own faults." He then repeats his views on dietary laws, the strictness of the Sabbath, and circumcision. To add insult to injury, he even adds a few lines in which he uncritically repeats the old antisemitic stereotype about a supposed Jewish commercial spirit!

When reviewing books by Jewish authors that do not discuss contemporary practices, Nöldeke rarely draws attention to their religious background. Occasionally, however, he comments on what he considers to be two major weaknesses in Jewish scholarship. One recurring criticism was that some Jewish authors stuck too close to "the authority of the old Jewish tradition."[52] As the preceding section has illustrated, however, Nöldeke did not reserve this reproach for Jewish authors. He occasionally criticized his Protestant and Catholic peers on the same grounds. Nöldeke observed a second major weakness in the writing style of his Jewish peers: "A flaw that is, alas, common among Jewish writers [...] is a too flowery account and a tendency to embellish in the depiction of Jewish events."[53] He notes that one Jewish author should use a more "prosaic" style and that another should have "cut out some pompous expressions."[54]

All in all, however, Nöldeke was not significantly more critical of the works of Jewish authors than of those by others. He wrote twenty-seven reviews of books by Jewish writers and four of these were unambiguously unfavourable, which makes for an unfavorability rating that hardly diverges from his general tendencies. It is also noteworthy that fifteen of these twenty-seven works were reviewed in the *Theology* section. So even if Nöldeke may have been somewhat more likely to criticize books by Jewish authors, this can also be explained by the fact that he received a disproportionally high number of books by Jewish authors about theological topics. And, as the preceding section has shown, Nöldeke tended to be more critical of theological treatises by pious followers of all religions.

Nöldeke's opinions of Catholicism were hardly more favourable than his thoughts on Judaism. When he was asked to act as an expert witness in an Austrian blood libel court case, he sarcastically commented on it by quoting a Heinrich Heine poem: "But it simply seems to me / That the rabbi and the monk / that both of them they stink."[55] Most of his critical comments about dogmatism were directed at Catholics rather than Jews. These critiques are most clearly summarized in the wry dismissal of Hermann Zschokke's discussion of the Old Testament: "The book may not contain a lot of independent judgements."[56] Most of Nöldeke's assessments of books by Catholic authors, however, were more favourable.

A lot of the works by Catholic authors that Nöldeke received for review were text editions and translations of the Syriac Church Fathers and their contemporaries. His most important complaint about such editions was not that they were incompetently executed, but that the selected texts were hardly worthy of publication. Catholic scholars, he argued, tended to select and organize their texts based on religious considerations. He criticized the grouping of texts in Gustav Bickell's *Conspectus Rei Syrorum*, for example, for being based on the church's assessment of whether they were "orthodox" or "heretic."[57] He characterized Bickell's selection for his collection of texts of Syriac Church Fathers as restricted "to such older Church writers, who he deems to be strictly orthodox."[58] Another collection of Syriac texts was also dismissed for its irrelevance: "The content of most pieces contained in this volume, is, alas, once again quite insignificant, at least in comparison to the many important things that the editor could have found among the Roman manuscripts."[59]

Notwithstanding his criticism, Nöldeke found that these largely insignificant text editions held at least some interest: they could provide a picture of early Christian thought. Even though he saw little merit in a new edition of the writings of Isaac of Antioch, he admitted that "they are still important as a document of the views, feelings, and desires of the Christian Syrians at a time, when these played a very significant role in the development of the church and its dogma's."[60] So if such an edition was competently edited, he would encourage his Protestant peers to buy these inconsequential products of Catholic scholarship anyway.[61] All in all, Nöldeke was about as likely to write a largely favourable review of the work of Catholic authors as of Jewish or Protestant scholars.

Finally, there is the question of Nöldeke's assessment of foreign scholarship. In private he was quite critical of French scholarship in his field: "It is a sad situation with the Arabists in Paris, anyway. Who holds the chairs of De Sacy and Quatremêre? Ever since Guyard died, Derenbourg junior has the whole field for himself. Zotenberg, who is superior to all of them, is pushed completely into the background."[62] He also repeatedly commented on the supposed vanity of the French.[63] However, even though he reviewed Sédillot's *Histoire Générale des Arabes* unfavourably, he was not particularly critical of other French books.[64] The other foreign books he reviewed were written by authors with different nationalities, such as British, Dutch, Italian, and Russian. In total, Nöldeke reviewed thirty-six foreign books and six of his reviews were unambiguously dismissive. This does not allow for any conclusions about possible prejudices towards non-German scholarship.

Wundt on Religion and Nationality

The high number of Jewish scholars working in Semitic languages and Old Testament studies was quite exceptional. In the medical faculty and the non-linguistic disciplines in the philosophical faculty, the Jewish presence was more limited. Alexandra Pawliczek's study of Jews appointed at the University of Berlin suggests that the Faculty of Law was most welcoming to Jews: more than a quarter of its staff was born in a Jewish family.[65] Almost 10 percent of the students with a Jewish background who finished the *Habilitation* eventually reached the position of full professor. The number of people with Jewish roots was even higher at the Faculty of Medicine but only 3 percent of the Jews who finished their *Habilitation* there would eventually become full professors. Jewish career perspectives at the Faculty of Philosophy were even more limited, even if the high percentage of Jewish scholars in oriental studies is taken into account.

In a way, the lack of career perspectives for people with a Jewish background was more problematic for students of medicine than for orientalists. The orientalists could not only work as *Privatdozenten* in the periphery of the academic system, but they could also find employment at a variety of Jewish religious and educational institutions.[66] The medical scholars reviewed by Wundt could not resort to such alternatives. They were even more dependent on university structures because physiological and anatomical research required workspaces, research materials, and tools that were only available at the well-endowed research institutes of German academia. At these places, Jews felt a strong pressure to convert to Christianity. Of all German professors with a Jewish background, only 13 percent had *not* been baptized.[67] Therefore, most of the authors with a Jewish background whose work was reviewed by Wundt no longer identified as Jewish. Scholars like Julius Bernstein, Jacob Henle, and Rudolf Heidenhain had grown up in Jewish families but identified as Protestants. And because religious background is not as obviously relevant to medical research as it is to Old Testament studies, it is hardly surprising that there is no discernible aversion of—or preference for—the work of Jewish authors in Wundt's reviews.

Similarly, there was no reason for Wundt to be particularly critical of the work of Catholic authors. A considerable part of the books by Catholic scholars that Wundt reviewed was written by Austrians and Viennese medical scholarship was held in high regard across Europe in the late nineteenth century. The renown of the Vienna Medical School can be traced back to the eighteenth century but reached new heights one century later in what has been called the new "Golden Age of medicine in Vienna."[68] The Vienna General Hospital led the way in pathological anatomy and clinicopathological studies.[69] Physiology was well-represented by Ernst Brücke, a German protestant who was repeatedly praised by Wundt.[70] In a review of a study of Sigmund Exner, Wundt acknowledges the exemplary character of Brücke's institute, at which the author was employed.[71] His review of the institute's leader's own *Vorlesungen über Physiologie* was also favourable: "Neither the professional physiologist nor the student will indeed put these lectures aside without have derived a lot of instruction from it."[72] In the same issue of the *Centralblatt*, he praises Brücke's colleague Joseph Hyrtl as well: "We have no doubts that [this research method] will receive more attention than it has received so far, thanks to this work by the famous Viennese anatomist."[73]

In this light, there is no compelling reason to assume that Wundt was more critical of either non-Germans or non-Protestants. Even though I have not been able to determine the nationality of all authors whose work he reviewed, I have been able to establish the national identity of most of them and the resulting overview suggests that he was equally critical of Germans and foreigners. At least 39 of Wundt's 123 reviews are of books by non-German authors. I have not been able to establish the nationality of four of the unfavourable reviewed authors, but they all have German-sounding names.[74] If we assume that of these people Jos Raith is Austrian, because his book was published in Vienna, it seems permissible to assume that the other three authors, whose books were published in Germany, are German. This would mean that Wundt reviewed thirteen out of eighty-two German and six out of forty foreign books dismissively, which amounts to a very similar ratio between favourable and unfavourable reviews for both groups.

Praising and Criticizing Books

One striking feature of the book reviews in the *Centralblatt* is that they reflected both on the qualities of the reviewed books and on those of their authors. This is a quality that is not unique to nineteenth-century book reviews: Hyland presents similar findings in his analysis of a corpus of hundreds of late twentieth-century book reviews.[75] In the next section, I will pay attention to the ways in which reviews were used to evaluate individual scholars and draw attention to shared conceptions of good scholarship. In this section, however, I will take a closer look at the features of the books themselves that were singled out for praise and criticism. This will not only result in an overview of the most common reasons for praise and criticism but will also provide the contours of shared conceptions of good scholarship and illustrate the way in which book reviews were used to establish and maintain such conceptions.

One feature that was often singled out for praise was the perceived importance of the reviewed work. One reason a book could be important was because it contained new findings. Nöldeke, for example, praises the al-Mokadassi edition of his friend De Goeje as "a book that increases our knowledge of the Orient in an excellent way."[76] Novelty was not the only reason a publication could be of interest, however. Ferdinand Wüstenfeld's edition of Abu Obaid Abd Allah Ibn Abd al-Aziz al-Bakri's geographical dictionary receives Nöldeke's praise for being "a most rich reference book about the localities found in the ancient Arabic poetry and in a part of the traditional literature."[77] In his favourable reviews, Nöldeke repeatedly uses the same evaluative terms to draw attention to the importance of the work, such as *wichtig* (important), *bedeutend* (significant), *reich* (rich), *wertvoll* (valuable), and *nützlich* (useful).

For Wundt, the criterions of novelty and general importance figured among the main reasons to write a positive review as well. He, for example, recommends a work on microscopy with the following words: "As a result of its inclusion of many completely new ways of experimenting, it will be a welcome addition for those who own one of the major works on microscopic technology."[78] A book on the anatomy of the brain was not praised for its novel insights, but is nonetheless considered to be of interest because of its clear presentation of the state of the art in its field: "[…] our literature does not yet possess a work that, in a similar way, provides a generally understandable […] exposition of the brain's anatomy."[79] In general, however, Wundt tended to give more favourable reviews of innovative works than of books providing an overview of the current consensus in a field of research.

The perceived importance of the reviewed work was not the only element evaluated by the reviewers: they also paid close attention to the manner of presentation. Nöldeke was particularly drawn to an attractive and clear style of writing. In his review of the second posthumous volume of Abraham Geiger's writings he notes that the writing is "fresh and warm" and praises "the humane undertone of [Geiger's] being that characterises all his work."[80] Reinhart Dozy, De Goeje's former teacher, is also complimented for his writing style: "That the book commends itself through its brilliant conception and glowing exposition, goes in Dozy's case without saying."[81] Terms that Nöldeke uses for this type of praise often include *klar* (clear), *anziehend* (attractive), *geschmackvoll* (tasteful), and *anregend* (stimulating).

Wundt also frequently reflects on the way in which research is presented. Usually, he does not comment on the authors' writing styles, however, but draws attention to illustrations instead. A publication about the larynx, which may not have been of interest to a very wide audience, is nonetheless praised because "it was illustrated with numerous and excellent woodcuts in such a way that it can indeed not be hard even for the anatomically and physiologically uneducated to obtain a quite extensive knowledge of the important organ."[82] At times, he also shows enthusiasm about those works that benefited from the use of the new medium of collotype, a technology that made it possible to print photographs.[83] He praises a book on the nerves in the hand by emphasizing that the "plates in collotype have turned out excellently and give a renewed proof of the beautiful enrichment, which the anatomic expositions has gained with the adaption of photography."[84] Late nineteenth-century collotype, however, was often too crude to depict delicate phenomena in great detail.[85] This is also made clear in another review by Wundt, in which he emphasizes that detailed depictions of the brain "can clearly still be reproduced more faithfully through copperplate and even through woodcut […] than by means of collotype."[86]

As I have noted above, both Nöldeke and Wundt tended to be most critical when they reviewed contributions about more abstract and contemplative topics. A disproportional share of Nöldeke's contributions to the *Theology* section is highly critical. This can partly be understood in the light of Suzanne Marchand's characterization of him as "firmly ensconced in Enlightenment rationalism."[87] This stance makes even more sense in the light of his uneasy relationship with his *Doktorvater*, Heinrich Ewald. Already in the 1860s Nöldeke distanced himself from the morally charged and theologically inspired studies of Ewald and cultivated his ties with the positivist text analysis advocated by Heinrich Leberecht Fleischer.[88] A similar distrust of dogma and speculation is noticeable in Wundt's reviews. He harshly criticized books grounded in *Naturphilosophie*, Fichtean and Schopenhauerian idealism, Hegelian dialectics, phrenology, and the belief in the veracity of wondrous observations of spiritists.

These observations about the elements that Nöldeke and Wundt selected for praise and critical scrutiny suggest that book reviews contributed in a meaningful way to the creation and maintenance of ideals—or even communities—of scholarship. The recurring reasons for praise, which were as diverse as significance, novelty, an attractive writing style, and insightful illustrations, identified the most basic requirements of good scholarship. When these requirements were fulfilled, this affirmed the reviewed author's membership of a peer group with shared ideals of excellence. The acknowledgment of the existence of a largely like-minded peer group, however, also draws attention to the fact that not every scholar was perceived to share the same standards. The critical way in which Nöldeke and Wundt discussed approaches to theology and philosophy that they considered to be dogmatic or speculative, provides a first indication that book reviews were also used to draw a line between an in-group and an out-group. In this way, their reviews allowed them to show loyalty towards one group of authors, while asserting their independence from others.

Praising and Criticizing Authors

Even though book reviews are ostensibly about books, they also contain judgements and expectations of these books' authors. In this section, I will investigate two ways in which book reviews tell us something about the desirable qualities, character traits, and virtues of scholars. First, I pay attention to the question to what extent these reviews were used to create and maintain a distinction between university educated and affiliated researchers on the one hand, and lay publicists on the other hand. Any obvious differentiation of this kind would suggest that reviews are indeed used to strengthen the cohesion of a strictly academic community of peers. Next, I investigate the terms used by reviewers to praise authors rather than their books. This provides an insight in the qualities that academics considered to be at the root of the distinction between university trained and affiliated researchers and outsiders with an interest in the same topics.

The vast majority of the books reviewed in the *Centralblatt* was written by men with a university affiliation. This was in line with Zarncke's aim to create a journal "that envisioned a comprehensive overview of scholarly literature" without "renouncing the strictest standards of our scholarship in any way."[89] In this light, it is noticeable that more than half of Nöldeke's twelve blatantly negative reviews discuss books by authors without an academic affiliation or even a doctorate. These authors constitute a rather colourful group of people: Bernard Neteler was the vicar of Loburg Castle, Joseph Samuel Bloch was some years away from finishing his doctorate in Zürich, Adolf Brüll taught at the *Philantropin*, a Jewish institution for secondary education in Frankfurt, Adolf Koch was a gymnasium professor in a small town in Switzerland, Giuseppe Barzilai was an attorney in Trieste, and John Mühleisen-Arnold was the rector of a church in Cape Town.[90] I could not, alas, find any information about Georg Janichs' occupation at the time of the publication of the new edition of his licentiate's thesis.[91]

Nöldeke's criticism of these authors without university affiliation suggests that he used his book reviews at least to some extent to draw a line between academic peers and lay outsiders. He emphasizes several times that the academically unaffiliated authors lack even the most basic academic skills. Barzilai is written off as a "dilettante," Janichs is accused of insufficient knowledge of Syriac, and Mühleisen-Arnold misses "the necessary knowledge for scholarly judgement of the Islam." His criticism of what he considered to be lay dilettantism is further amplified by the religious themes of the critically assessed texts. As I have argued above, Nöldeke tended to write off works that he considered to be religiously dogmatic. He applied this criticism to the lay scholarship of the vicar Neteler, the teacher at a Jewish high school Brüll, and the church rector Mühleisen-Arnold as well. About the latter he, for example, stated that "his theological point of view and his theological bias do not allow him an unprejudiced judgement of a non-Christian religion at all."[92]

Wundt's pattern of criticism of authors without a university affiliation is strikingly similar to Nöldeke's. The group of lay authors reviewed by Nöldeke largely consisted of authors of theological treatises, of which he was critical to begin with. The group of unaffiliated authors evaluated by Wundt largely consists of people who wrote about philosophical topics: a field that brought out the fierce critic in Wundt. The critically reviewed authors can be divided into three main categories: those who advocated new but—in Wundt's eyes—unpromising fads, those advocating old-fashioned philosophies, and a small group of authors accused of misrepresenting Kant. The first group largely consisted of spiritists, such as the Russian journalist and translator Alexander Aksakov. In a review of a German journal edited by Aksakov, Wundt stated with relief that it proved that spiritism was "an exotic growth in Germany [...] that does not truly prosper among us."[93] The attorney Robert Schellwien was one of the authors dismissed because of his advocacy of a supposedly obsolete philosophy: *Naturphilosophie*.[94] The jurist and politician Julius von Kirchmann, finally, was one of the authors criticized for misrepresenting Kant and offering their own idiosyncratic interpretation of his philosophical legacy instead.[95] Other people's interpretations of Kant were important to Wundt because he valued his work both as a barrier against the most high-minded claims of idealist philosophies and as a rebuttal of what he saw as the crudeness of nineteenth-century positivism.[96]

The large majority of Wundt's reviews, however, was of works by authors who had a position as *Ordinarius* or *Extraordinarius* at a university. Most of these reviews were neutrally descriptive or explicitly positive. He gave a negative evaluation in only five reviews of works by authors with a university appointment beyond the starting position of *Privatdozent*. Only one of these reviews was of a book reviewed for the medical section of the *Centralblatt*: William Preyer's work on Weber's law, a topic very close to Wundt's heart.[97] Another of his critical reviews was of a book by the same author that was discussed in the *Natural Sciences* section. The book presented a philosophical theory of the four elements, which was ridiculous in Wundt's eyes.[98] The other few negative reviews of professorial peers were all published in the philosophical section and discussed the work of proponents of philosophical traditions that Wundt considered to be outdated. All in all, however, it appears that Wundt either highly appreciated the work of established academics or felt that it was only rarely appropriate to sharply criticize them.

While Wundt's favourable reviews of established scholars are notable, in Nöldeke's case his reviews of promising new members of the academic peer group are particularly remarkable. All published dissertations merited words of praise, with the odd exception of one book, in which the author argued that Muhammed's pride and lewdness were the cause of his rejection of the Christian faith.[99] The other reviews of dissertations typically offered their authors a warm welcome into academic circles. One review concluded with the words: “We hope that we can meet the author again as a contributor to the field of oriental studies.”[100] Another stated that he expected “quite some contributions to scholarship of this young scholar, who can already present such a competent accomplishment.”[101] His review of the dissertation of one of De Goeje’s most promising students, Theodoor Houtsma, concluded with the statement that he was “looking forward with high expectations to [the author’s] promised investigations about the further development of the Islamic dogmatic theology.”[102] These reviews unambiguously signalled that Nöldeke was willing to accept the new doctors as members of a scholarly community.

Finally, it is interesting to take a quick look at the vocabulary used to praise authors rather than books. In addition to the terms used to recommend scholarly works that I discussed in the previous section, there existed a series of recurring compliments for their authors as well. The most common virtues for which authors were praised in Nöldeke’s reviews were *Fleiss* (industriousness), *Sorgsamkeit* (diligence), *Umsicht* (cautiousness), *Gelehrsamkeit* (erudition), *Scharfsinn* (ingenuity), and *Kritik* (a critical stance). In Wundt’s reviews, the virtues of *Fleiss* and *Umsicht* are repeatedly mentioned as well. In addition, he strongly emphasized the importance of experimentation and empiricism. This partly overlapped with Nöldeke’s praise for *Umsicht* and *Scharfsinn*: “the experiments of these researchers seem to have been conducted with great caution and partly with the use of a very ingenious technique.”[103] This interest in experimental methods also nicely dovetails with Wundt’s aversion of speculative philosophies: “The author does not deduce final pictures of structural coherence from his observations; we can be all the more sure that the trustworthiness of the latter does not suffer from the influence of hypotheses made in advance.”[104]

The observations in this section illustrate the ways in which book reviews dealt not only with books but also with authors in at least two ways. In the first place, we have seen that reviewers created a vocabulary that could be used to criticize or praise authors rather than their work. In addition, we have found that complimentary book reviews were used to distinguish a scholarly community of academically affiliated peers from a wide variety of lay scholars who published about the same topics. Both Nöldeke and Wundt often criticized outsiders. Their reviews of their university colleagues tended to be much more favourable. A closer look at these favourable reviews shows that there were some differences in the exact way in which Nöldeke and Wundt distributed their praise. Wundt’s reviews show that he was unlikely to be highly critical of his professorial peers. Nöldeke's reviews, however, show that they could also be used to welcome recently promoted doctors into an academic community.

Addressing a Community of Scholars

The creation and maintenance of scholarly communities is a familiar topic in the history of scholarship and the sociology of scientific knowledge. Steven Shapin has, for example, drawn attention to scientists' perceptions of their communities and the structural virtues that characterize them, such as universalism and disinterestedness.[105] Thomas Gieryn has discussed the way in which scientific communities attempt to distinguish themselves from communities that they consider to be unscientific.[106] A closer look at the way in which shared virtues and demarcation criteria are presented in book reviews illustrates how scholars contributed to such debates about a scientific identity through everyday practices of scholarship.

Book reviews have a twofold audience consisting of both the author whose work is reviewed and a wider community of peers. Because of this double audience, reviewed authors find themselves in a vulnerable position. After all, they are not just judged, they are judged in front of their peers. Reviewers find themselves in a somewhat uncomfortable position as well. Even though they are free to show "little deference" to the reviewed authors, their relation to their wider audience is more complicated.[107] Hyland states that reviewers can pose a "threat to the wider community by adopting a position of authority in relation to it."[108] Others, however, do not stress this type of threat, but instead emphasize that reviewers are expected to show a certain humility towards the wider community of scholars instead.[109] The reviewer also has to consider the relation between the reviewee and the peer group. A book review can be seen as a face-threatening act for the reviewee precisely because it is shared with an audience of his or her peers. The different hopes and expectations that different people have of these different audiences call for "a *polite* realisation of critical remarks" in order to maintain a "proper balance between collegiality and critique."[110] A successful exercise of politeness strategies in book reviewing is therefore essential to the maintenance of communities of scholarship.

Such strategies are described at length in a seminal study by Penelope Brown and Stephen C. Levinson. Brown and Levinson distinguish positive and negative politeness strategies. Positive politeness strategies involve satisfying the audience's desire for recognition "by communicating that one's own wants (or some of them) are in some respects similar to the addressee's wants."[111] Negative politeness strategies consist of efforts to "modify the force of a speech act" or—in other words—attempts to provide potentially face-threatening criticism in a sympathetic, non-judgemental manner.[112] As I will show in the following section, both Nöldeke and Wundt make use of positive as well as negative politeness strategies in order to soften the blows of critical observations and to underline and sustain solidarity among academics who disagree about all kinds of scholarly topics.

One common positive politeness strategy, that is primarily directed at reviewed authors, is what Brown and Levinson call "gift giving" which can include the gifting of "sympathy" and "understanding."[113] All the different kinds of praise for books and authors discussed in the earlier sections fall into this category. Another, less explicit, common politeness strategy is assuring one's readers that author, reviewer, and reader all share the same interest. In one of his reviews Nöldeke, for example, explicitly addresses "all friends of sound interpretations of the O[ld] T[estament]."[114] In another review, he softens his criticism by assuring that "there will not be a lot of readers [...] who will be bothered by the mentioned shortcoming."[115] Wundt likewise postulates shared interests when he states that "nobody will hesitate to acknowledge that a certain advance is possible and desirable in this direction."[116] In another review, he emphasizes that the book "will be used with benefit by those readers who have some educational background in the natural sciences.[117]" Such phrasings explicitly present the reviewer, the reviewee, and other readers as members of a scholarly community with shared aims. When Wundt states that "we have indeed always lacked a textbook of this kind until now," the word "we" refers to himself, the book's author Jacob Henle, and all of the readers interested in the anatomy of the hand.[118]

Another positive politeness strategy, that Brown and Levinson discuss, but that Nöldeke and Wundt rarely resort to, is joking.[119] The few examples of humour in their reviews are rather an invitation to the readers of the *Centralblatt* to share a laugh about an oddly unsatisfying book than an effort to appease the reviewee. Nöldeke's mocking remark about needing "much less subjection of reason to believe in the 'authenticity' of the Pentateuch [...] than to believe in the infallibility of the pope" is an example of such an attempt at somewhat mean-spirited humour.[120] One extremely short review, by Wundt, that I quote here in full, displays a similar biting irony:

> The conclusion of this work is contained in the proposition that logic is an a posteriori science. If the arguments of the author would have been as substantial as this proposition, which he calmly expresses, the opus at hand would be an epoch-making publication. But because these arguments are basically limited to the well-known possibility of the geometrical conception of logical relationships, we can suffice with this short note.[121]

Neither Nöldeke nor Wundt, however, mastered the humorous review as well as Gustav Theodor Fechner. Under the moniker of Dr Mises, he had earlier published satirical essays in which he had proved, among other things, that the moon was made of iodine, and engaged in a comparative anatomy of angels.[122] With this background, he was eminently suited to poke fun at a poetic analysis of the relationship between the soul and the body:

> We would like to ask the author only this one question […], why, while otherwise […] paying attention to all details in his depiction of the human form, and especially the human face, he has overlooked the nose, which, in our opinion, hardly deserves this poetic neglect; a flaw that will, by the way, easily be repaired in a second edition […]. Just imagine […]; a face without a nose![123]

In the accompanying letter to Zarncke, he claimed to have "constructed the review in such a way that […], at best, a light wholesome doubt might arise in [the author], whether he is being treated ironically." Realizing that the possibility existed that the victim of his wisecracks would not take them as lightly as he hoped he would, he added that he would not protest, however, if Zarncke would decide to delete his facetious comments about noseless faces.[124]

In addition to these instances of positive politeness strategies, the reviews of Nöldeke and Wundt also contain many instances of negative politeness strategies. Such strategies are usually exercised in book reviews in order to save the reviewee's face in front of an audience of his peers. They often take the shape of hedges that soften the impact of otherwise severe criticism. One frequent way in which both Nöldeke and Wundt were able to tone down their criticism was by insisting that their critiques drew attention to minor issues in otherwise important and well-executed studies. Wundt, for example, argued that even though he could think of a few criticisms of minor details in a concise book about consciousness, "nobody will put down the lucidly and appealingly written little book without feeling very inspired."[125] In one of his book reviews, Nöldeke qualified his own criticism by emphasizing that he "only has to deviate from [the author] on a few trivialities."[126]

Another negative politeness strategy consisted of stressing that the tasks that the reviewed authors had set for themselves were so ambitious that it would have been impossible to complete them without some minor shortcomings. This strategy can be interpreted as an example of what Brown and Levinson call "relevance hedges," which are based on the assumption that criticism loses some of its edge in the light of the other merits of the reviewed book.[127] Ambition and novelty were certainly seen as such merits. Even in his largely negative review of Georg Janichs' book, Nöldeke, for example, admits that the topic of his research was "extremely difficult, in part," thereby suggesting that his critique did not take anything away from the overall importance of the author's efforts.[128] In one review, Wundt likewise draws attention to the fact that physiology is a very challenging field of research because it "finds itself in continuous transformations."[129] Despite his criticism, he therefore recommended the reviewed book to all scholars with at least a basic knowledge of the latest developments in its field.

A final common negative politeness strategy consists of drawing attention to one's own lack of relevant specialist knowledge. This strategy is very similar to what Brown and Levinson call "quality hedges," which suggest "that the speaker is not taking full responsibility for the truth of his utterance."[130] After all, even the most critical review loses some of its sting when it is written by someone who admits that he may have misunderstood or overlooked certain valuable qualities of the reviewed work. Wundt rarely opts for this strategy, but Nöldeke often points at gaps in his own knowledge. In a review of a book about a Christian Syriac text, he draws attention to the fact that he is "alas not well-versed [...] in liturgical issues," which means that that he "has to declare himself incompetent to judge exactly those excerpts, to which the author attaches the most importance."[131] In his commentary on a text edition that draws on Coptic sources, he admits that the "final judgment of their critical value" should be left to "the experts of the Coptic language," of which he was not one.[132]

The politeness strategies used by Nöldeke and Wundt contributed to the creation and maintenance of scholarly communities in at least two ways. The negative politeness strategies allowed them to criticize books and authors, while at the same time assuring that the reviewees would not lose face in front of a wide audience of their peers. As such, these strategies acknowledged the dual nature of the book reviews' audience and validated the reviewees' desire to be accepted into a community of scholars. The bond between reviewer, reviewee, and peer group was even more explicitly emphasized through positive politeness strategies. While the strategy of gift-giving was exclusively directed to the authors of the reviewed works, frequent reminders of the shared interests of reviewer, reviewee, and audience gave expression to a sense of solidarity and common values among the peer group as a whole.

Loyalty and Independence in Book Reviews

Fritz Ringer has noted that late nineteenth-century German academics shared a well-defined sense of self. They generally agreed that "the modern German idea of the university and of learning was irrevocably tied to its intellectual origins in German Idealism and neohumanism."[133] The ideal of the educated man as an "autonomous personality" can be traced back to the ideals of classical *Bildung* that shaped German gymnasiums at least from the late eighteenth century onwards. This appreciation of autonomy and individuality was an unmistakable feature of the religious ideals of *Kulturprotestantismus* that advocated a religiosity that went together surprisingly well with "a laicist persuasion" and a strong interest in the empirical and rational pursuit of scholarly research.[134] This encouraged even scholars who identified as Protestants to assert their independence from dogmatic religious attitudes and speculative philosophical claims.

The book reviews of Nöldeke and Wundt show that this self-image of independent critics of dogma and speculation lay at the heart of their understanding of good scholarship as well. It was particularly urgent to them to protect the critical character of theology and philosophy. In Nöldeke's eyes, the scientific character of Old Testament studies was threatened by the uncritical, dogmatic contributions of orthodox Protestant, Catholic, and Jewish authors. Wundt was a reliable critic of almost all advocates of *Naturphilosophie*, idealists of the Hegelian, Fichtean, and Schopenhauerian persuasion, spiritists, and other philosophers he considered to be too speculative. These assertions of intellectual independence divided scholars into an outgroup of speculative dogmatists and an ingroup of empirical rationalists. In this way, the ideals of scholarship that convinced Nöldeke and Wundt to distance themselves from certain colleagues also contributed to a shared conception of a community of scholars with shared values.

Nöldeke and Wundt believed that the lofty values of autonomous scholarship and detachment from dogma and mysticism were best served by the scholarly virtues that they repeatedly singled out for praise. False religious certitudes and the deceptions of speculation should be counteracted by the virtues of caution, industriousness, diligence, and ingenuity. The laboratory offered the perfect space for these virtues to blossom. Ingeniously conducted experiments, careful measurements, and diligently conducted series of personal observations were seen as the perfect means to distinguish between the domain of scholarship and the realms of superstitious speculation. Only such careful, empirical studies could receive the coveted praise of being significant, important, valuable, or useful. Only reviewed works that embodied these virtues were even considered for additional praise for the lucidity of their exposition. Nöldeke complimented authors with an agreeable and clear writing style and Wundt commented favourably on state-of-the-art and elucidating illustrations.

The distinction between uncritical dogmatists and autonomous scholars was applied both to discriminate between scholars with different intellectual backgrounds and to protect the boundary between academic and lay scholarship. Abstract and speculative disciplines like theology and philosophy were particularly likely to be populated with lay authors. The study of these topics was, after all, not dependent on the access to well-preserved, rare manuscripts or the access to well-equipped clinics or laboratories. Therefore, authors without a university affiliation were disproportionally likely to be reviewed unfavourably both because of their field of study and on the basis of their apparent lack of academic credentials.

The maintenance of a critical distance from an unprofessional, dogmatic, or speculative outgroup also created the contours of ingroup of supposedly rational, careful, diligent, and ingenious scholars. And while the book review was a medium through which independence from an outgroup could be asserted, it was also a medium that allowed reviewers to display their loyalty to the reviewees and their scholarly readership at large. This becomes apparent both in Wundt's tendency to either write neutral or favourable reviews of his professorial peers and in Nöldeke's use of book reviews to welcome talented, young scholars to a wider scholarly community.

Loyalty between reviewer, reviewee, and readers is further encouraged through a wide array of politeness strategies used by both Nöldeke and Wundt. They were constantly aware that they addressed not only an author in their book reviews. They realized that their reviews were read by a broad audience of their peers. They also recognized that it was exactly the existence of this wide audience that turned their reviews into face threatening acts for the reviewees. Therefore, they made careful use of positive and negative politeness strategies in their reviews. These strategies included explicit expressions of solidarity and understanding, the acknowledgement of their own lack of knowledge, jokes, emphasizing the importance of even imperfect scholarship, and highlighting the insignificance of their criticism in the light of the scope and ambition of the reviewed work.

The use of such politeness strategies is itself an acknowledgement of their appreciation of the close relation between reviewer, reviewee, and a broader audience. Furthermore, because the repeated, explicit emphasizing of shared aims and interests was an important element of these strategies, the reviews often amounted to explicit acknowledgements of the existence of a community with a shared conception of scholarly virtue. Therefore, book reviews were uniquely suited to communicate that diligence, clarity of exposition, carefulness, and caution were and should be valued by everyone who wanted to be taken seriously as a scholar. In addition, the intellectual modesty shown by the sympathetically polite reviewer was yet another quality that members of the peer group could be expected to display.

In this intellectual environment, ambitious but careful readers of book reviews learned that they could be accepted as part of the scholarly community if they wrote important, novel books in a clear and attractive style. These same careful readers also learned that they were expected to project an image of diligence, caution, ingenuity, and modest politeness. In living up to these expectations they could not only find acceptance in academic circles, but they would also affirm and reinforce the self-image of a community of scholarship based on exactly these requirements.

In the earlier chapters, loyalty and independence were largely presented as virtues that are most likely to surface in the relation between individuals. The second chapter presented a close look at the private correspondence between individual scholars and the third chapter mainly—though certainly not exclusively—drew attention to the relation between the editors of scholarly journals and their contributors and other stakeholders. The example of the *Zeitschrift der Deutschen Morgenländischen Gesellschaft*, however, already pointed at a way in which loyalty to individuals could be intertwined with the loyalty to institutions and the peer community. The processes of ingroup and outgroup creation and maintenance in this chapter add another part to the complex puzzle of nineteenth-century scholarly loyalties. In the next chapter on university appointments and letters of recommendation even more types of performances of loyalty and independence will be discussed.

Notes

1. Michael Jan de Goeje to Theodor Nöldeke, 21 April 1867, Universiteitsbibliotheek Leiden (UBL): BPL 2389.
2. Theodor Nöldeke to Michael Jan de Goeje, 12 September 1867, UBL: BPL 2389.
3. Ute Schneider, "Die Funktion wissenschaftlicher Rezensionszeitschriften im Kommunikationsprozess der Gelehrten," in *Kultur der Kommunikation: Die europäische Gelehrtenrepublik im Zeitalter von Leibniz und Lessing*, ed. Ulrich Johannes Schneider (Wiesbaden: Harrassowitz, 2005), 281–282.
4. Miriam Nicoli, "Faced with the flood: scholarly working practices and editorial transformations at the highpoint of scientific publication," in *Scholars in action: the practice of knowledge and the figure of the savant in the 18th century*, ed. Andre Holenstein, Huberts Steinke and Martin Stuber (Leiden: Brill, 2013), 610.
5. Ute Schneider, "Die Funktion wissenschaftlicher Rezensionszeitschriften," 283.
6. Thomas Habel, *Gelehrte Journale und Zeitungen der Aufklärung: Zur Entstehung, Entwicklung und Erschließung deutschsprachiger Rezensionszeitschriften des 18. Jahrhunderts* (Bremen: Edition Lumière, 2007), 222.
7. Johann Christoph Greiling, "Einige allgemeine Grundsätze zu einer Theorie der Recensionen," *Archiv für die Physiologie*, Dritter Band (1799): 353.
8. Ken Hyland, *Disciplinary Discourses: Social Interaction in Academic Writing* (Ann Arbor, MI: University of Michigan Press, 2004), 44.
9. Rosa Lorés-Sanz, "Local Disciplines, Local Cultures: Praise and Criticism in British and Spanish History Book Reviews," *Brno Studies in English* 38, no. 2 (2012): 98.
10. Anthony Giddens, "Durkheim as a Review Critic," *The Sociological Review*, New Series 18, no. 2 (1970): 171.

11. Philipp Müller, "Geschichte machen. Überlegungen zu lokal-spezifischen Praktiken in der Geschichtswissenschaft und ihrer epistemischen Bedeutung im 19. Jahrhunder. Ein Literaturbericht," *Historische Anthropologie* 12, no. 3 (2004): 430.
12. [Friedrich Zarncke], untitled editorial introduction, *Literarisches Centralblatt* 1 (1850): 1.
13. Friedrich Zarncke, "An unsere Leser," *Literarisches Centralblatt* 52 (1874), 1–2.
14. Wilhelm Wundt, "Rubinstein, Dr. Sus., die sensoriellen und sensitiven Sinne," *Literarisches Centralblatt* 22 (1875): 22.
15. This analysis will mostly draw on: Penelope Brown and Stephen C. Levinson, *Politeness: Some universals in language usage* (Cambridge: Cambridge University Press, 1987); Ken Hyland, *Disciplinary Discourses*; Greg Myers, "The pragmatics of politeness in scientific articles,' *Applied Linguistics* 10, no. 1 (1989).
16. Thomas Lick, *Friedrich Zarncke und das 'Literarische Centralblatt für Deutschland.' Eine buchgeschichtliche Untersuchung* (Wiesbaden: Harrassowitz, 1993), 33–35.
17. In making these distinctions, I follow Claudia Profos Frick, who even makes more fine-grained distinctions between very negative, negative, undecided, positive, and very positive reviews: Claudia Profos Frick, *Gelehrte Kritik: Albrecht von Hallers literarisch-wissenschaftliche Rezensionen in den Göttingischen Gelehrten Anzeigen* (Basel: Schwabe, 2009): 268–287, 299. A similar classification of criticism in reviews can be found in Françoise Salager-Meyer, María Ángeles Alcaraz Ariza and Maryelis Pabón Berbesi, "Collegiality, critique and the construction of scientific argumentation in medical book reviews: a diachronic approach," *Journal of Pragmatics* 39 (2007): 1762ff.
18. Theodor Nöldeke, "Barzilai, Dr. G., le lettere dell'alfabeto fenicio," *Literarisches Centralblatt* 30 (1876): 987.
19. Rudi Paret, *Arabistik und Islamkunde an deutschen Universitäten: Deutsche Orientalisten seit Theodor Nöldeke* (Wiesbaden: Franz Steiner, 1966), 14.
20. Theodor Nöldeke, "Zschokke, Dr. Herm., Prof., historia sacra Antiqui Testamenti," *Literarisches Centralblatt* 1 (1873): 1–2.
21. Theodor Nöldeke, "Neteler, B., Studien über die Echtheit des Pentateuchs," *Literarisches Centralblatt* 37 (1873): 1.
22. Theodor Nöldeke, "Böhl, Ed., Forschungen nach einer Volksbibel zur Zeit Jesu," *Literarisches Centralblatt* 37 (1873): 1153–1157.
23. Theodor Nöldeke, ""Fürst, Jul., Geschichte der biblischen LIteratur und des jüdisch-hellenistischen Schriftthums," *Literarisches Centralblatt* 14 (1871): 337–339.

24. Wilhelm Wundt, "Preyer, W., das myopische Gesetz," *Literarisches Centralblatt* 32 (1874): 1039–1041. Wundt's emphasis.
25. Wilhelm Wundt, "Büchner, Dr. Ludw., physiologische Bilder," *Literarisches Centralblatt* (49): 1576.
26. Wilhelm Wundt, "Ranke, Dr. Joh., Prof., Grundzüge der Physiologie des Menschen," *Literatisches Centralblatt* 21 (1873): 651–653.
27. Wilhelm Wundt, "Rauch, P.M., die Einheit des Menschengeschlechtes," *Literarisches Centralblatt* 12 (1873): 358–359.
28. Wilhelm Wundt, "Volkelt, Dr. Joh., Die Traum-Phantasie," *Literarisches Centralblatt* 31 (1876): 1001–1003.
29. Wilhelm Wundt, "Schellwien, Rob., das Gesetz der Causalität in der Natur," *Literarisches Centralblatt* 33 (1877): 1091–1092; Wilhelm Wundt, "Entleutner, A.F., Naturwissenschaft, Naturphilosophie und Philosophie der Liebe," *Literarisches Centralblatt* 52 (1877): 1713–1714.
30. Wilhelm Wundt, "Preyer, Wilh., über die Erforschung des Lebens," *Literarisches Centralblatt* 25 (1873): 775.
31. Wilhelm Wundt, "Schmitz-Dumont, Zeit und Raum," *Literarisches Centralblatt* 30 (1876): 975–976; Wilhelm Wundt, "Meynert, Th., Prof., zur Mechanik des Gehirnbaues," *Literarisches Centralblatt* 5 (1875): 136–137.
32. Wilhelm Wundt, "Michelet, C.L., das System der Philosophie als exacter Wissenschaft," *Literarisches Centralblatt* 9 (1877): 269–270.
33. See, Chap. 2, 43.
34. Wilhelm Wundt, "Owen, R.D., (1) das streitige Land (2) Psychische Studien," *Literatisches Centralblatt* 21 (1877): 687–689.
35. Wilhelm Wundt, "Carneri, B., Gefühl, Bewußtwein, Wille. Eine psychologische Studie," *Literarisches Centralblatt* 9 (1877): 268–269.
36. Wilhelm Wundt, "Volkmann v. Volkmar, Dr. Wilh., Prof., Lehrbuch der Psychologie," *Literarisches Centralblatt* 2 (1876): 33–34.
37. Wilhelm Wundt, "Baumgärtner, Heinr., die Weltzellen," *Literarisches Centralblatt* 18 (1876): 591–592.
38. See this chapter, 116–122.
39. Constantin Goschler, "Deutsche Naturwissenschaft und naturwissenschaftliche Deutsche: Rudolf Virchow und die 'deutsche Wissenschaft'," in *Wissenschaft und Nation in der Europäischen Geschichte*, ed. Ralph Jessen and Jakob Vogel (Frankfurt: Campus, 2009), 110–111.
40. William Clark, *Academic Charisma and the Origins of the Research University* (Chicago, IL: University of Chicago Press, 2006), 28.
41. Anders Gerdmar, *Roots of Theological Anti-Semitism: German Biblical Interpretation and the Jews, From Herder and Semler to Kittel and Bultmann* (Leiden: Brill, 2009), 3.

42. Robert Irwin, *For Lust of Knowing: The Orientalists and their Enemies* (London: Penguin, 2007), 198.
43. Suzanne L. Marchand, *German Orientalism in the Age of Empire: Religion, Race, and Scholarship* (Cambridge: Cambridge University Press, 2009), 77.
44. Aleksandra Pawliczek, "Kontinuität des informellen Konsens: Die Berufungspolitik der Universität Berlin und ihre Dozenten im Kaiserreich und in der Weimarer Republik," in *Kontinuitäten und Diskontinuitäten in der Wissenschaftsgeschichte des 20. Jahrhunderts*, ed. Rüdiger von Bruch, Uta Gerhard tand Aleksandra Pawliczek (Stuttgart: Franz Steiner, 2006), 70.
45. Ari Joskowicz, *The Modernity of Others: Jewish Anti-Catholicism in Germany and France* (Stanford, CA: Stanford University Press, 2014), 5.
46. Marchand, *German Orientalism*, 77.
47. Johannes Renger, "Altorientalistik und jüdische Gelehrte in Deutschland—Deutsche und österreichische Altorientalisten in Exil," in *Jüdische Intellektuelle und die Philologien in Deutschland 1871–1933*, ed. Wilfried Barner and Christoph König (Göttingen: Wallstein, 2001), 249.
48. Ignaz Goldziher would be one of the few exceptions. In 1905 he was appointed to full professor in Budapest. His career was strongly shaped by struggle with antisemitic prejudice, however. See, for example: Peter Haber, *Zwischen jüdischer Tradition und WissenschaftL Der ungarische Orientalist Ignác Goldziher (1850–1921)* (Cologne: Böhlau, 2006).
49. Theodor Nöldeke to Michael Jan de Goeje, 9 November 1874, UBL: BPL 2389.
50. Theodor Nöldeke, "Geiger's, Abrah., nachgelassene Schriften," *Literarisches Centralblatt* 32 (1875): 1048–1050.
51. Theodor Nöldeke, "Meyer, S., ein Wort an Herrn Hermann Messner," *Literarisches Centralblatt* 52 (1877): 1711–1712.
52. Theodor Nöldeke "Bloch, J.H., Ursprung und Entstehungszeit der Buches Kohelet," *Literarisches Centralblatt* 12 (1873): 353–354.
53. Theodor Nöldeke, "Gross, Dr. Siegm., Menahem ben Saruk," *Literarisches Centralblatt* 28 (1872): 743–744.
54. Theodor Nöldeke, "Cassel, Dr. David, Geschichte der jüdischen Literatur," *Literarisches Centralblatt* 12 (1873): 353–354; Theodor Nöldeke, "Masechet Soferim. Der talmudische Tractat der Schreiber," *Literarisches Centralblatt* 16 (1879): 505–506.
55. Theodor Nöldeke to Michael Jan de Goeje, UBL: BPL 2389. In German, it reads: 'doch es will mich schier bedünken daß der Rabbi und der Mönch daß sie alle beide stinken." The English translation is taken from: Renate Schlesier, "Homeric Laughter by the Rivers of Babylon: Heinrich

Heine and Karl Marx," in *The Jewish Reception of Heinrich Heine*, ed. Mark H. Gelber (Tübingen: Niemeyer, 1992), 36.
56. Theodor Nöldeke, "Zschokke, Dr. Herm., Prof., historia sacra Antiqui Testamenti," *Literarisches Centralblatt* 1 (1873): 1–2.
57. Theodor Nöldeke, "Bickell, Gustavus, conspectus rei Syrorum," *Literarisches Centralblatt* 30 (1871): 756–758.
58. Theodor Nöldeke, "Ausgewählte Schriften der syrischen Kirchenväter übersetzt von Prof. Dr. Gust. Bickell," *Literarisches Centralblatt* 12 (1877): 371–372.
59. Theodor Nöldeke, "Monumenta syriaca et Romanis codicibus collecta," *Literarisches Centralblatt* 29 (1878): 953–955.
60. Theodor Nöldeke, "J. Issaci Antiocheni," *Literarisches Centralblatt* 8 (1873): 225–228.
61. Theodor Nöldeke, "Ausgewählte Gedichte der syrischen Kirchenväter," *Literarisches Centralblatt* 28 (1872): 725–727.
62. Theodor Nöldeke to Michael Jan de Goeje, 3 August 1886, UBL: BPL 2389.
63. Theodor Nöldeke to Michael Jan de Goeje, 27 November 1858, UBL: BPL 2389; Theodor Nöldeke to Michael Jan de Goeje, 15 September 1859, UBL: BPL 2389.
64. Theodor Nöldeke, "Sédillot, L.-A., Prof., histoire générale des Arabes," *Literarisches Centralblatt* 35 (1877): 1172.
65. Pawliczek, "Kontinuität des informellen Konsens," 73–75.
66. Céline Trautmann-Waller, "Selbstorgainsation jüdischer Gelehrsamkeit und die Universität seit der 'Wissenschaft des Judentums'," in *Jüdische Intellektuelle und die Philologien in Deutschland, 1871–1933*, ed. Wilfried Barner and Christoph König (Göttingen: Wallstein, 2001).
67. Notker Hammerstein, "Universitäten in Kaiserreich und Weimarer Republik und der Antisemitismus," in *Jüdische Intellektuelle und die Philologien in Deutschland, 1871–1933*, ed. Wilfried Barner and Christoph König (Göttingen: Wallstein, 2001), 30.
68. Alfred Vogl, "Six hundred years of medicine in Vienna: A history of the Vienna School of Medicine," *Bulletin of the New York Academy of Medicine* 43, no. 4 (1967): 290–291; Eric Kandel, *The Age of insight: The Quest to Understand the Unconscious in Art, Mind, and Brain, from Vienna 1900 to the Present* (New York: Random House, 2012), 22–23.
69. Udo Schagen, "Germany, Austria, and Switzerland," in *The Oxford Companion to Medicine*, ed. Stephen Lock, John M. Last and George Dunea (Oxford: Oxford University Press, 2006), retrieved from https://www.oxfordreference.com/view/10.1093/acref/9780192629500.001.0001/acref-9780192629500-e-208.

70. For more on Brücke's background in the Prussian academic world, see: Gerald Wiest and Robert W. Baloh, "The Personal and Scientific Feud Between Ernst Brücke and Josef Hyrtl," *Otoloty & Neurotology* 27 (2006): 570–571.
71. Wilhelm Wundt, "Exner, Dr. Sigm., Privatdoc., Leitfaden bei de mikroskopischen Untersuchung thierischer Gewebe," *Literarisches Centralblatt* 26 (1873): 813.
72. Wilhelm Wundt, "Brücke, Ernst, Vorlesungen über Physiologie," *Literarisches Centralblatt* 39 (1873): 1229.
73. Wilhelm Wundt, "Hyrtl, Dr. Jos., Prof., die Corrosions-Anatomie, *Literarisches Centralblatt* 39 (1873): 1228–1229.
74. Wilhelm Wundt, "Schmitz-Dumont, Zeit und Raum," *Literarisches Centralblatt* 30 (1876): 975–976; Wilhelm Wundt, "Raith, Jos., Entdeckungen im GEbiete der geistigen Verrichtungen des Centralnervenssystems," *Literarisches Centralblatt* 10 (1877): 298; Wilhelm Wundt, "Krause, Alb., die Gesetze des menschlichen Herzens," *Literarisches Centralblatt* 14 (1877): 454; Wilhelm Wundt, "Entleutner, A.F., Naturwissenschaft, Naturphilosophie und Philosophie der Liebe," *Literarisches Centralblatt* 52 (1877): 1713–1714.
75. Hyland, *Disciplinary Discourses*, 47.
76. Theodor Nöldeke, "Descriptio imperii Moslemici," *Literarisches Centralblatt* 28 (1877): 923–926.
77. Theodor Nöldeke, "Das geographische Wörterbuch," *Literarisches Centralblatt* 52 (1877): 1727–1728.
78. Wilhelm Wundt, "Exner, Dr. Sigm., Privatdoc., Leitfaden bei de mikroskopischen Untersuchung thierischer Gewebe," *Literarisches Centralblatt* 26 (1873): 813.
79. Wilhelm Wundt, "Luys, J., Arzt, das Gehirn, sein Bau und seine Verrichtungen," *Literarisches Centralblatt* 1 (1878): 12.
80. Theodor Nöldeke, "Geiger's, Abrah., nachgelassene Schriften," *Literarisches Centralblatt* 32 (1875): 1048–1050.
81. Theodor Nöldeke, "Dozy, R., Essai sur l'histoire de l'Islamisme," *Literarisches Centralblatt* 13 (1879): 406–407.
82. Wilhelm Wundt, "Merkel, Dr. C. Ludw., der Kehlkopf," *Literarisches Centralblatt* 32 (1873): 1003.
83. On collotypes, see *The Grove Encyclopedia of Materials and Techniques in Art*, ed. Gerald W.R. Ward (Oxford: Oxford University Press, 2008), 113.
84. Wilhelm Wundt, "Braune, Wilh. und A Trüdiger, die Venen der menschlichen Hand," *Literarisches Centralblatt* 19 (1874): 630.
85. Martin Kemp, "Medicine in View: Art and Visual Representation," in *Western Medicine: An Illustrated History*, ed. Irvine Loudon (Oxford: Oxford University Press, 1997), 6.

86. Wilhelm Wundt, "Rüdinger, Dr. u. Prof., topographisch-chirurgische Anatomie des Menschen," *Literarisches Centralblatt* 37 (1874): 1234–1235.
87. Marchand, *German Orientalism*, 176.
88. Christiaan Engberts, "Gossiping about the Buddha of Göttingen: Heinrich Ewald as an Unscholarly Persona," *History of Humanities* 1, no. 2 (2016): 378.
89. Friedrich Zarncke, "An unsere Leser," *Literarisches Centralblatt* 52 (1874): 1722–1723.
90. Theodor Nöldeke, "Neteler, B., Studien über die Echtheit des Pentateuchs," *Literarisches Centralblatt* 1 (1873): 1; Theodor Nöldeke, "Bloch, J.H., Ursprung und Entsehungszeit der Buches Kohelet," *Literarisches Centralblatt* 12 (1873): 353–354; Theodor Nöldeke, "Brüll, Dr. Ad., das samaritanische Targum zum Pentateuch," *Literarisches Centralblatt* 17 (1874): 553–555; Theodor Nöldeke, "Koch, Ad. Prof., Moabitisch oder Selimisch?,"*Literarisches Centralblatt* 13 (1876): 425–428; Theodor Nöldeke, "Barzilai, Dr. G., le lettere dell' alfabeto fenicio," *Literarisches Centralblatt* 30 (1876): 987; Theodor Nöldeke, "Arnold, John Mühleisen, Lic. der Theol. u. Dr. ph., der Islam nach Geschichte, Charakter u. Beziehung zum Christenthum," *Literarisches Centralblatt* 35 (1878): 1148.
91. Theodor Nöldeke, "Janichs, Dr. Georg, Lic. Theol., animadversions criticae in versionem Syriacam Peschitthonianam, *Literarisches Centralblatt* 49 (1871): 1225–1227.
92. Theodor Nöldeke, "Arnold, John Mühleisen," 1148.
93. Wilhelm Wundt, "Owen, R.D., (1) das streitige Land (2) Psychische Studien," *Literarisches Centralblatt* 21 (1877): 687–689.
94. Wilhelm Wundt, "Schellwien, Rob., das Gesetz der Causalität in der Natur," *Literarisches Centralblatt* 33 (1877): 1091–1092.
95. Wilhelm Wundt, "Kirchmann, J.H. v., Erläuterungen zu Kant's Schriften zur Naturphilosophie," *Literarisches Centralblatt* 24 (1878): 786–787.
96. Wilhelm Wundt, *Erlebtes und Erkanntes* (Stuttgart: Alfred Kröner, 1920), 56. See also: Edwin G. Boring, *A History of Experimental Psychology*, second edition (Englewood Cliffs, NJ: Prentice-Hall, 1950), 246–250; Daniel N. Robinson, *An Intellectual History of Psychology*, revised edition (New York, NY: Macmillan, 1981), 325–326.
97. Wilhelm Wundt, "Preyer, W., das myopische Gesetz," *Literarisches Centralblatt* 32 (1874): 1039–1041.
98. Wilhelm Wundt, "Preyer, Wilh., über die Erforschung des Lebens," *Literarisches Centralblatt* 25 (1873): 775–776.
99. Theodor Nöldeke, "Bethge, Frideric., Raḥmân et Aḥmad," *Literarisches Centralblatt* 26 (1872): 671–673.

100. Theodor Nöldeke, "Sasse, C.J. Franz, Prolegomena in Aphrastis Sapientis Persae sermones homilecticos," *Literarisches Centralblatt* 13 (1879): 401–402.
101. Theodor Nöldeke, "Fünf Streitfragen der Basrenser und Kufenser [...] herausgegeben und bearbeitet von Dr. Jaromir Košut," *Literarisches Centralblatt* 38 (1878): 1271–1273.
102. Theodor Nöldeke, "Houtsma, M.Th., de strijd over het dogma," *Literarisches Centralblatt* 34 (1875): 1106–1108.
103. Wilhelm Wundt, "Arbeiten aus der physiologischen Anstalt zu Leipzig," *Literarisches Centralblatt* 12 (1873): 357–358.
104. Wilhelm Wundt, "Stieda, Dr. Ludw., Prof., über den Bau des centralen Nervensystems der Amphibien und Reptilien," *Literarisches Centralblatt* 2 (1876): 44.
105. Steven Shapin, *The Scientific Life, A Moral History of a Late Modern Vocation* (Chicago, IL: Chicago University Press, 2008), 15–16.
106. Thomas F. Gieryn, "Boundary-Work and the Demarcation of Science from Non-Science: Strains and Interests in Professional Ideologies of Scientists," *American Sociological Review* 48 (1983): 781.
107. Myers, "The pragmatics of politeness," 4.
108. Hyland, *Disciplinary Discourses*, 45.
109. Myers, "The pragmatics of politeness," 4.
110. Salager-Meyer, Alcaraz Ariza and Pabón Berbesi, "Collegiality, critique and the construction of scientific argumentation,"1771. My emphasis.
111. Brown and Levinson, *Politeness*, 101.
112. Ibid., 145.
113. Ibid., 129.
114. Theodor Nöldeke, "Tuch, Wil. Dr. Fr., Commentar über die Genesis," *Literarisches Centralblatt* 24 (1871): 601–602.
115. Theodor Nöldeke, "Fürst, Jul., Geschichte der biblischen Literatur und des jüdisch-hellenistischen Schriftthums," *Literarisches Centralblatt* 14 (1871): 337–338.
116. Wilhelm Wundt, "Aeby, Chr., der Bau des menschlichen Körpers," *Literarisches Centralblatt* 8 (1872): 186–187.
117. Wilhelm Wundt, "Huxley, Thom. H., Grundzüge der Physiologie," *Literarisches Centralblatt* 33 (1872): 883–884.
118. Wilhelm Wundt, "Henle, J., anatomischer Hand-Atlas," *Literarisches Centralblatt* 5 (1872): 106.
119. Brown and Levinson, *Politeness*, 124.
120. Theodor Nöldeke, "Neteler, B., Studien über die Echtheit des Pentateuchs," *Literarisches Centralblatt* 37 (1873): 1.

121. Willhelm Wundt, "Widemann, Paul, über die Bedingungen der Uebereinstimmung des discursiven Erkennens mit dem intuitiven," *Literarisches Centralblatt* 10 (1877): 298.
122. Hans-Jürgen Arendt, "Gustav Theodor Fechner (1801–1887) und die Leipziger bürgerliche Gesellschaft im 19. Jahrhundert," *N.T.M. Zeitschrift für Geschichte der Wissenschaften, Technik und Medizin* 9 (2001): 5.
123. [Gustav Theodor Fechner], Hauschild, Ed. Ferd., Psyche, oder der Becher Giamschid's," *Literarisches Centralblatt* 26 (1853): 427–428.
124. Gustav Theodor Fechner to Friedrich Zarncke, no date, UBLE: NL 249/1/F/61.
125. Wilhelm Wundt, "Carneri, B., Gefühl, Bewußtsein, Wille. Eine psychologische Studie," *Literarisches Centralblatt* 9 (1877): 268–269.
126. Theodor Nöldeke, "Smend, Rud., Moses apud prophetas," *Literarisches Centralblatt* 42 (1875): 1345–1346.
127. Brown and Levinson, *Politeness*, 168.
128. Theodor Nöldeke, "Janichs, Dr. Georg, Lic. Theol., animadversions criticae in versionem Syriacam Peschitthonianam," *Literarisches Centralblatt* 49 (1871): 1225–1227.
129. Wilhelm Wundt, "Huxley, Thom. H., Grundzüge der Physiologie," *Literarisches Centralblatt* 33 (1872): 883–884.
130. Brown and Levinson, *Politeness*, 164.
131. Theodor Nöldeke, "Bickell, Gustavus, conspectus rei Syrorum," *Literarisches Centralblatt* 30 (1871): 756–758.
132. Theodor Nöldeke, "Psalterium Job. Proverbia arabice, Paulus de Lagarde edidit," *Literarisches Centralblatt* 2 (1879): 33–35.
133. Fritz K. Ringer, *The Decline of the German Mandarins: The German Academic Community, 1890–1933* (Cambridge, MA: Harvard University Press, 1969), 103–104.
134. Gangolf Hübinger, *Kulturprotestantismus und Politik: Zum Verhältniss von Liberalismus und Protestantismus im wilhelmischen Deutschland* (Tübingen: Mohr, 1994), 15.

CHAPTER 5

State and Scholarship: The Politics of Professorial Appointments

Peer Review and Grant Review

The term "peer review" is often used to refer to the editorial practice of consulting external experts before deciding on the acceptance or rejection of a paper.[1] The preceding chapter has drawn attention, however, to the fact that scholars do not limit themselves to assessing each other's work: they also judge each other's personal and professional qualities.[2] One field in which this type of peer review is obviously important is the evaluation of grant proposals. The literature of grant review panels is limited.[3] The lack of studies on such assessments before the nineteenth century also suggests that this type of peer review was not very common until quite recently. Only in the second half of the nineteenth century, some precursors of grant review panels have become a somewhat prominent feature of academic life.

One of the earlier examples can be found in the United Kingdom. From 1850 onwards, the Royal Society of London was charged with distributing some government funds for scientific research. This led to the establishment of the *Committee of Recommendations*, which would be known as the *Government Grant Committee* from 1853 onwards. In the first year of its existence, seven applications by fellows of the Royal Society were accepted and six applications by non-fellows were rejected.[4] The exact criteria for the allotment of grants cannot be reconstructed but a clear preference existed for "the collecting of facts" by fellows of

© The Author(s), under exclusive license to Springer Nature Switzerland AG 2022

C. Engberts, *Scholarly Virtues in Nineteenth-Century Sciences and Humanities*, https://doi.org/10.1007/978-3-030-84566-7_5

the society.[5] A similar development took place in France. Before the mid-nineteenth century, the *Académie Royale des Sciences* almost exclusively rewarded scholarship that had already been carried out with symbolic prizes, such as honorary medals. In the second half of the nineteenth century, however, an increasing amount of money was spent on the funding of future research. It is, however, hard to get a clear view of the criteria for such funding because these decisions were hardly given any publicity and some grants were even awarded behind closed doors.[6]

Eighteenth-century German academies worked in a similar way as their French counterparts: they shaped scholarship through prize contests rather than by funding research proposals.[7] In the nineteenth century, however, the *Königlich-Preußische Akademie der Wissenschaften* started to fund the research initiatives of some of its members, as well. Financial support was not typically awarded based on the assessment of different research proposals. Instead, the *Akademie* decided to fund a small number of highly ambitious, grand projects initiated by influential scholars, like August Böckh's *Corpus Inscriptionum Graecarum* and Theodor Mommsen's *Corpus Inscriptionem Latinarum*.[8] Even when the *Akademie* eventually decided to also award grants to individual scholars in the second half of the nineteenth century, it continued to favour ambitious projects and the support of prestigious research institutes, such as the *Deutsches Historisches Institut* in Rome, the *Geodätisches Institut* in Potsdam, and the *Meteorologisches Institut* in Berlin.[9]

These modest innovations in the funding of research in places like London, Paris, and Berlin might be the closest thing to a modern-day grant review panel in the late nineteenth century. Their practices do not teach us a lot about the way in which individual scholars judged and supported each other. The committees of the Royal Society seem to have cared primarily about the applicants' membership of the organization, the Parisians hid their decision-making process from the public eye, and the Germans were mostly interested in grand projects and prestigious institutes. A better understanding of the way in which individual scholars were able to influence each other's careers through the assessment of each other's past work, personal qualities, and career prospects should therefore be based on a closer look at another pervasive evaluative practice: university appointment procedures. These often involved a high number of letters of recommendation that are particularly suited to shed light on the ways in which assertions of loyal collegiality and critical independence shaped scholars' careers and reflected their self-image.

The Althoff System

In nineteenth-century Germany, decisions about professorial appointments were made by state governments. Appointments at the University of Leipzig had to be approved by the Saxon government in Dresden, those at the University of Heidelberg by the Baden authorities in Karlsruhe, and those at Prussian universities by the Ministry of Education in Berlin. The decision-making process in Berlin provides a particularly promising starting point for further research for at least two reasons. In the first place, the relation between the Prussian state and its universities provided a blueprint for the whole of imperial Germany.[10] In the second place, for a quarter of a century, this process was largely controlled by one man, whose correspondence has been very well preserved. This correspondence offers a close look into Prussian appointment politic during a long period.

This one man in charge of these affairs was Friedrich Theodor Althoff. His official position in the ministerial hierarchy was quite modest but his influence should not be underestimated. Between 1882 and 1897, he was one of the thirty-three *vortragende Räte* (executive officials) at the Ministry of Education and between 1897 and 1907 he was one of its four *Ministerialdirektoren* (ministerial directors).[11] His influence was disproportionate, however, as the way in which contemporary news media commented on him illustrates. They referred to him as "the almighty ruler of the Prussian universities," "The secret Prussian Minister of Education," and "the Bismarck of the university system."[12] It has become common to refer to his legacy as the "Althoff system." The word "system," however, says more about the perceived inescapability and decisiveness of his influence than about the clarity or coherence of his vision. The modern-day chronicler of the Althoff system, Bernhard vom Brocke, emphasizes that he was "a practical man, not at all doctrinaire" and draws attention to the fact that his system "slowly emerged in the course of decades of educational policy-making."[13]

Nonetheless, Althoff's policies show several common threads. One noted feature was an unwillingness to discriminate against minorities.[14] He proudly claimed that he never "participated in any rabble-rousing propaganda, neither against Catholics nor against Jews."[15] Another characteristic of his approach was a strong interest in research institutes that functioned outside of the universities' faculty structure.[16] His support of Robert Koch's Institute for Infectious Diseases is only one typical example of this. Another feature that characterized Althoff's rule was his ability to find new ways to fund research. The budget of his ministry was limited, but he successfully forged bonds between academia and private and corporate investors.[17] Finally, if his approach can be described as a system, it is best characterized as a system for gathering information.[18] At every Prussian faculty, Althoff cultivated relations with at least one informant, who would inform him of the accomplishments, ambitions, and character traits of its members. This allowed him to make well-informed decisions about professorial appointments and to influence decision-making processes at the faculty level.

Of course, not everybody was pleased with Althoff's seeming omnipotence. Especially his skilful bypassing of the existing power structures as manifested in the universities' faculty structure was widely criticized. Traditionally, professorial appointments were the product of mutual consultation between a faculty and the Ministry. The faculty would send a ranked list of preferred candidates to the Ministry. Even if the minister was not obliged to follow the faculty's preferences, he would usually appoint one of the listed candidates.[19] Althoff's ability to shape careers through the support of research institutes existing outside of faculty structures therefore challenged traditional academic power structures. In addition, his welcoming attitude towards Catholics and Jews was disliked by the often more conservative faculty members.[20] This aversion of Althoff could also turn into a critique of his informants. The historian of mathematics David Rowe argues that Althoff's *Vertrauensleute* were viewed "with great suspicion" and that some even "thought of them as unscrupulous informants, practically akin to academic spies."[21] The harshest criticism, however, was reserved for Althoff himself. The philosopher Max Dessoir even suggested a relation between the Althoff system and the powerlessness of German academics towards the Nazis a few decades later: "We forgot how to speak and to act like free men; even after Althoff's death, we continued to live in 'fear of the Lord' and readily changed to another line of policy when National Socialism began to 'coordinate' us."[22]

Althoff's critics may have overstated his power, or his willingness to use it. An analysis of the number of professorial appointments of scholars not listed by the faculty shows that such appointments were relatively uncommon during his years of tenure.[23] His warm relationship with a remarkably high number of influential scholars allowed him to shape opinions within the faculty without having to fall back on the full force of his administrative powers. Occasionally, Althoff even gave in to well-organized resistance within a faculty. The Faculty of Philosophy in Greifswald, for example, successfully advocated for its own candidate for the Chair of Classical Philology. Initially, Althoff was not very enthusiastic about the young classicist Eduard Norden, but because the faculty was unanimous in its resistance against the mediocre older scholar preferred by Althoff, he eventually relented.[24]

The Althoff system is not my main interest in this study: it is only the context within which the performances of loyalty and independence, in which I am primarily interested, occurred. In this chapter, I will take a closer look at the way in which scholars used their confidential correspondence with the powerful *Ministerialdirektor* to evaluate their colleagues in the full knowledge that this evaluation could decisively shape their peers' careers. The next section will illustrate how Althoff acquired in-depth knowledge of both the professional and personal qualities of scholars by looking at his intimate knowledge of the events and interpersonal relationships at the *Seminar für Orientalische Sprachen* (Institute for Oriental Languages) in Berlin. The following sections show how scholars at the medical and philosophical faculties used their connections to Althoff to shape both the careers of individual scholars and the character of their faculties. This will provide an overview of the way in which university policy advice and letters of recommendation enabled scholars to both loyally support some of their peers and to critically distance themselves from others. In addition, this chapter will paint a picture of what professional and personal qualities were considered to be most relevant to a successful academic career in late nineteenth-century Germany.

Althoff's Intimate Knowledge

The Institute for Oriental Languages was founded in 1887, in the wake of the young sinologist Wilhelm Grube's warnings about the lacking fluency in local languages at the German embassy in China. He pointed out that the French and British governments had established institutes with the explicit aim to educate future diplomats and businessmen in Asian vernaculars. In Germany, however, no such institutions existed.[25] Grube proposed the founding of an institute modelled on the French *École des langues orientales vivantes*, where European teaching staff was supported by native speakers. Only a few days after the Minister of Education, Gustav von Goßler, had pitched the idea to Bismarck, it was approved.[26] Althoff was entrusted with the preparations, and two years later the Institute for Oriental Languages opened its doors under the direction of Eduard

Sachau, a former student of Nöldeke and professor of Arabic at the University of Berlin. From these early days onwards, the relationships between the institute's staff members were strained and nobody knew more about the simmering tensions than Althoff.

Althoff corresponded extensively with several staff members of the institute, such as Sachau and the Arabist Martin Hartmann. Even though the writer of one of his obituaries would later argue that he had "absolutely no talent for gossip," the sinologist Carl Arendt turned out to be one of Althoff's most trusted and useful informants.[27] In 1889, he asked him for a detailed confidential report on the work atmosphere at the institute. Arendt repeatedly emphasized that he saw this request as a "painful and probably questionable assignment" but nonetheless painted an almost juicy and certainly worrying picture of the work climate at the institute.[28] He praised some staff members, such as the Chinese lectors working under his supervision, but he harshly criticized a lot of his other colleagues. Strikingly, even though he explicitly praised his great working relationship with Sachau, he saved some of his most biting criticism for the institute's director.

Arendt mostly complained that Sachau was unapproachable. He was only occasionally present at the institute and almost all communication between him and the other staff members was through written missives. His office hours were scheduled at a time when no other staff members had any reason to be at the institute and requests for private meetings were usually turned down. Because Sachau hardly cared for the institute's daily routines, his planning ignored the everyday experiences of the teaching staff. All complaints about this state of affairs were brushed aside as a "revolt against his authority" or "impermissible criticism of his actions." What made matters even worse, was that Sachau did not treat his staff members equally. A translation request was, for example, politely worded to all professors, but one of them received it "in the shape of a decree bordering between business-like and almost rude." Especially Martin Hartmann and the Persianist Friedrich Rosen felt threatened by the institute director. His dislike of them was so obvious that even the institute's students noticed it!

Arendt's report also drew attention to Sachau's own confidants at the institute. He was particularly close to the lecturer in Arabic Bernhard Moritz and the institute attendant (*Seminardiener*) Wilhelm Heyde. Whenever Sachau did not communicate by means of written missives, his directives were usually passed on by these men. The knowledge that he was the director's favourite had turned Moritz into an insufferable colleague. He saw himself as "the mouthpiece of the director and occasionally even as [his colleague's] superior." His comments about Arendt's teaching and general rudeness defied "the most common forms of courtesy." Heyde's place in the institute hierarchy was officially lower than that of the professors and lecturers. It raised some eyebrows, therefore, when he was the one to announce the dismissal of Friedrich Rosen and Carl Friedrich Andreas, the professor of Turkish and Persian Languages. It was also through his indiscretion that the students were able to figure out the informal pecking order at the institute. Occasionally, Heyde would even be quite explicit in his rudeness towards the professors. When one of them did something that Sachau would not have approved, he only reacted with an ominous: "It is getting windy."[29]

Althoff's correspondence with other members of the institute confirms Arendt's depiction of a poisoned work environment. Hartmann mostly blamed Sachau and Moritz for Heyde's indiscretions.[30] Heyde had once tried to resist some of their inappropriate directives, but they had rebuffed him and told him that "he was cowardly" and that "he was too good." Hartmann therefore argued that "the improprieties and ineptitudes of attendant Heyde can be fully traced back to improprieties and ineptitudes of director Sachau." Even more damning incriminations of Sachau and Moritz reached Althoff's office from sources outside the institute. The Prussian consul in Damascus, Johann Gottfried Wetzstein, endorsed the accusations of plagiarism that Carl Friedrich Andreas directed at Sachau.[31] A former co-student of Moritz at the Berlin Faculty of Theology shared the story that he had only specialized in New Testament studies because Moritz had once stolen his lecture notes of the introductory course to the Old Testament.[32] All these incriminations were carefully filed away by Althoff.

It is not clear when and how Althoff used all this private information about the staff of the Institute for Oriental Language. Their careers do not suggest that he had any obvious preference for one scholar over the other. Sachau was never relieved of his directorial responsibilities; Arendt would keep his job at the institute until his death in 1902: Hartmann also stayed until his death in 1918; Rosen continued his career outside of the institute as a diplomat; Andreas spent two decades as a freelance language teacher in Berlin before he was hired by the University of Göttingen; Moritz, finally, left the institute for a position at Cairo's Khedival Library in 1896 and returned to the institute fifteen years later to succeed Sachau as its director.

This quick look at the Institute for Oriental Languages shows two things. First, it illustrates the degree to which Althoff was able to maintain confidential relationships with strongly opposing parties. Second, it shows how much information he obtained from these contacts. He kept himself updated not only about the scholarly merit of the institute's staff members but also learned about their personal character, their oftentimes strained relationships, and their occasionally inconvenient past and current missteps.

A Messy Breslau Faculty

The Institute for Oriental Languages was one of those typical Althoff inventions: a university institute that existed independent of the university's faculty structure. Therefore, we cannot assume that the correspondence between the institute's staff members and the *Ministerialdirektor* is also representative of the relation between the Ministry of Education and the still powerful faculties. One way in which the institute differed from university faculties was that its only task was language education, while university professors were expected to both teach and do research. Within some faculties the expectations were even higher. Medical faculties, for example, also had their own clinics and the professors working at those faculties could therefore be expected to carry out additional managerial tasks. In this section, I will look into the relation between Althoff and the Breslau Faculty of Medicine in the early 1890s. The Breslau case is particularly interesting because the faculty was struggling throughout the early 1890s. The continuous worries about the quality of its teaching and the management of its hospitals created an environment in which individual scholars had to balance their individual loyalties and independence and led to an awkwardly dependent relation with the Berlin authorities as well.

This investigation of the Faculty of Medicine is largely based on the correspondence between Althoff and the gynaecologist Heinrich Fritsch, his most trusted confidant in Breslau. Fritsch was professor of Obstetrics, director of the obstetrical clinic, and the administrative director of the faculty. His letters to Althoff show that this accumulation of managerial tasks during tumultuous times continuously threatened to overburden him: "I can basically give up on this year. If you have to argue every day about shirts, washing machines, meat deliveries, bickering by officials, etc, then where is there room for scientific research? […] But I am not going to waste another word on it, because I hope that I will not be staying in Breslau for much longer."[33] The reports that Fritsch sent to Berlin make his desire to leave quite understandable, indeed.

The most urgent threat was the lack of patients in the university clinic. This was the sad legacy of an earlier time, when Anton Biermer managed the medical clinic and the polyclinic. His rude attitude towards patients scared them away. Biermer, however, cherished the resulting reduction of his managerial tasks and was more than happy to work with the patients of the *Allerheiligen* hospital instead.[34] When Biermer fell ill in 1890, his clinical responsibilities were divided between Otto Soltmann and Friedrich Müller. Soltmann took over the medical clinic and Friedrich Müller was appointed as acting director of the polyclinic.[35] Soltmann and Müller managed their clinic in very different ways. Soltmann was a man of independent means, who hardly put any effort into the clinic and preferred to enjoy a copious lifestyle unbothered by teaching, research, or managerial responsibilities.[36] Müller, however, was an ambitious and hardworking man and his effective management of the polyclinic lured potential patients away from Soltmann's disorganized medical clinic.[37]

The successful Müller soon found a new job at the University of Marburg. By the time Alfred Kast arrived in Marburg as Biermer's successor, both clinics had only very few patients left. In his capacity as the faculty's administrative director, the well-connected Fritsch was burdened with the task to find new patients. Eventually, he reached an agreement with the local authorities that some of the poorer patients in the municipal medical facilities would be transferred to the university clinics: "Without this, we embarrass ourselves terribly with the beautiful medical clinic—without patients. We are not saving any money, because every patient is now hospitalized for free, so that we at least have *something*."[38]

The Faculty of Medicine in Breslau did not only lack patients in its clinics: it also had to deal with a dangerously low number of students. A worried Fritsch worked it out for Althoff. The only Prussian clinic with a lower student attendance than his obstetrical clinic was in Giessen, where only thirty-five students were enrolled. The 50 students registered at the Breslau clinic compared poorly, however, to the numbers of other small and mid-sized universities: Kiel had 88 students, Halle had 140 attendants, and Würzburg welcomed as many as 150 students.[39] One explanation for the students' unwillingness to visit the Breslau clinics, Fritsch argued, could be found in the unappealing and old-fashioned teaching and examination methods of the anatomist Carl Hasse. Even if he was "a good, consistent, and honest man" and "full of diligence and conscientiousness," his teaching was poor. He was more at home in animal than human anatomy and was an unnecessarily harsh examiner, whose shortcomings even made it into a student song:

Bummellied	Strolling Song
[...]	[...]
Schon 14 Tage vor Beginn	Already 14 days before the opening
Begann er sein College	He started his lecture series
Da schimpfte er, als ich mich einst	There he ranted when I
Verspätet auf dem Wege.	Was once delayed on my way.
Von Fröschen und Batrachiern war	Almost always the only things discussed
Fast immer nur die Rede.	Were frogs and batrachians.
Am Ende kam er etwas dann	In the end he would then come
Zum Menschen—ziemlich späte	To people—rather late.
[...]	
Nie nannte er uns Herr und Sie	He never called us gentlemen
Nur Er und Ihr und Leute	Just he and you and people
Im Mittelalter ging sowas	That was fine in the Middle Ages
Doch wahrlich nicht mehr heute.	But really no longer today.
Da müsst' ich doch ein Esel sein	I would really be a jackass
Ein Kerl als wie ein Rinde	A guy like a cow
Wenn ich nicht schleunig fahren sollt'	If I would not immediately go
Hinweg mit gutem Winde.	Away with a good wind.[40]

The song ruthlessly exposed Hasse's unpopular antics and its alarming consequences: an increasing number of Breslau students went "*hinweg mit gutem Winde*" and took their exams elsewhere. Leipzig was especially popular: many of those who could afford it chose to attend lectures there. This left Breslau with fewer and poorer students, which could make the university less appealing to students from wealthy and middle-class families. Fritsch distress is evident: "*There is no doubt that the attendance here will decline quickly*. The sons of *respectable* families already say: we cannot expose ourselves to Hasse's treatment and they go away. And then, only the rubbish stays with us!"[41]

Fritsch's loyalty was not primarily to the impolite Biermer, the uninterested Soltmann, the departed Müller, the unfortunate Kast, or the unloved Hasse but to the Breslau Faculty of Medicine itself. Since there was no easy way to deal with the problems caused by Hasse's teaching, Fritsch's recommendations were targeted at the clinics. He asked Althoff to abstain from appointing a successor for Müller because this could only encourage unhealthy competition between the struggling clinics.[42] Since this was an easy and cheap request to fulfil, it was quickly granted. When Fritsch's wish to leave Breslau was finally granted in 1893—he was appointed in Bonn—Althoff consulted him about a new professor in Obstetrics and manager of the Breslau obstetrical clinic as well. Even though Fritsch had entrusted both the medical clinic and the polyclinic to Kast, he recognized the need to appoint an actual obstetrician to manage his own clinic. His recommendation for this position was, again, not guided by personal loyalty but purely based on the perceived needs of the struggling faculty that he had left behind.

Fritsch advised Althoff to appoint Otto Küstner as his successor. His advocacy of Küstner had little to do with his scholarly accomplishments but was almost exclusively based on his personal wealth. Fritsch argued that "an independent man is necessary here. If you appoint a praxis-hungry professor from a small university, nothing will change, which [would be] disastrous for Breslau in particular."[43] The challenges of managing the struggling obstetrical clinic in Breslau required the full-time involvement of a director, who would not be distracted by any perceived need to add to his income by holding private practice as well. An additional advantage of appointing a wealthy man, Fritsch added, could be that he might easily be convinced to abstain from a travel and moving reimbursement and might even be willing to start his new job before receiving his first payment. Three months later, Althoff confirmed Küstner's appointment.[44]

Fritsch correspondence with Althoff reveals two things. In the first place, it shows that Althoff was not only briefed about the accomplishments and personal character of a wide range of German scholars, but that he was also kept up to date about the—occasionally worrying—states of affairs at universities, faculties, and associated institutions, such as university clinics. Secondly, it illustrates that loyalty to such institutions could be more important to his informants than loyalty to their individual colleagues. Even if Fritsch was likely to know scholars with the intellectual and personal qualities necessary to succeed Müller, the unsatisfying state of the university clinics prevented him from recommending any potential successor at all. In a similar vein, Fritsch must have been familiar with other obstetricians who were as qualified as, or more qualified than, Küstner to succeed him. His assessment that the Breslau clinic would benefit most from a director of independent means, however, tipped the balance towards recommending Küstner.

A Saxon Paediatrician

Of course, Althoff did not follow Fritsch's—or anyone's—advice blindly. In case of doubt, he double-checked his recommendation with other informants. The decision to hire Küstner was made easier by the fact that Hermann Kuhnt and Bernhard Schultze agreed with Fritsch's assessment of him.[45] It was not uncommon, however, for Althoff to receive contradictory evaluations. This happened when he was looking for the successor for the famous Berlin paediatrician Eduard Heinrich Henoch. One of the most promising candidates was the Leipzig researcher and clinician Otto Heubner, who we have already encountered as one of Emil Behring's main collaborators in the development of the diphtheria blood serum.[46] Some of his colleagues praised Heubner's work and character enthusiastically, but Eduard Külz—a "medical authority" and one of Althoff's most trusted informants—casted serious doubts on his suitability.[47]

The first person to whom Althoff turned, however, was Fritsch, and he did not even mention Heubner. In his short reflection on potential candidates, Fritsch mostly underlined the unfitness of Soltmann. He, again, drew attention to his preference for a very copious lifestyle. In reaction to Althoff's question whether Soltmann might have a Jewish background, Fritsch also painted a picture of him as a petty anti-Semite: "Soltmann is not Jewish, whether his father was Jewish, I do not know. Nonetheless, he looks very Jewish. This makes his antisemitism even more ridiculous. He owns a house in the most expensive neighbourhood of Breslau. Until today, he leaves two floors empty at a worth of over 1000 Thalers, because he does not want any Jews in the house."[48]

Hugo Falkenheim, a clinician and *Privatdozent* of paediatrics in Königsberg, was the first to bring Heubner to Althoff's attention. He recommended him as an "outstandingly diligent and solid acknowledged scholar whose works […] brought significant expansion of our knowledge."[49] He would be a worthy candidate because he was "the most generally distinguished of the German paediatricians; a man who qualifies himself to be Henoch's successor by his eminent expertise, his warm interest for paediatrics, and through his qualities as a teacher and a person." The physiologist Ludolf Krehl supported Falkenheim's praise in a long letter.[50] He drew attention to the quality and popularity of Heubner's lectures and his skilful management of both the polyclinic and the paediatric clinic. He also praised his personality: Heubner was "full of zest for life and not embittered" even though the Leipzig administrators had failed to reward his accomplishments with a well-deserved promotion for years.

Külz, however, referred to this career setback not to praise Heubner's character but to raise doubt about his suitability: "I would like to strongly emphasize the fact that [the Leipzig Faculty of Medicine] has passed him over for taking over the polyclinical professorship. Some people may say that his appointment did not have a chance in Dresden, but, of course, that cannot be the true and only reason."[51] Külz suggested that the slow career of Heubner was due to his typically Saxon personality: "His really too pronounced Saxon dialect may be the reason why his lectures, which are not bad, have always made me feel a bit funny. [...]." He therefore concluded that "[n]othwithstanding all the appreciation for some of his works, Heubner is a chatterbox. His comical personality comes across as ridiculous to his audience. It would be a great mistake to appoint him."[52] Külz could even draw on Kehl's otherwise favourable assessment of Heubner's lectures: "[He] is a very lively, typically Saxon figure; he speaks a strong Saxon dialect, which sometimes sounds funny. [...] His lectures suffer from [his] Saxon dialect; his way of talking is clumsy."[53] Külz's doubts were also supported by the influential Berlin pathologist Rudolf Virchow, who harboured similar doubts about Heubner.[54]

For Althoff, the decision was far from easy. Some of his informants wholeheartedly recommended Heubner as the foremost German paediatrician, while others argued that his Saxon manner of speaking made him unfit for a professorship in the Prussian capital. Most of the other candidates, however, were also far from perfect. Fritsch recommended Adolf Baginsky, Owald Kohts, and Carl von Noorden.[55] His praise for Baginsky, however, was ambiguous at best: "Baginsky is a Jew who forces himself to the foreground, but he certainly is not stupid." Appointing a Jew became even more unfeasible when the queen mother informed the ministry that she would prefer the main Berlin paediatrician—who might one day treat her children or grandchildren—to be a Christian.[56]

Fritsch did not even bother to dwell on his recommendation of Kohts and none of the other correspondents showed much enthusiasm for him either. Külz thought he was neither particularly hard-working nor very knowledgeable.[57] Even though he eventually concluded that Kohts would be the best candidate, his recommendation was hardly convincing: "There are favourable and unfavourable assessments of Kohts. Without further ado, I am more inclined to trust the unfavourable judgements. [...] [But] all the representatives of the discipline are inferior [to Henoch] without exception."[58] Falkenheim did not agree with Külz's choice for Kohts and pointed out that both Heubner and Baginsky were better qualified.[59] Van Noorden and Falkenheim's other preferred candidate, Oswald Vierordt, were not actually paediatricians, so Althoff never seriously considered appointing them. The final candidate, who was mentioned more than once, was Theodor Escherich. However, Falkenheim had added him to a list of people that should *not* be considered and Külz described him as "an overachiever" who "still has to mature."[60]

The position of the Berlin Faculty of Medicine is not clear. Some sources state that its members followed Virchow's lead and kept Heubner off their shortlist.[61] Others argue that both Henoch and the faculty supported Heubner's candidacy.[62] Faced with all these dissenting opinions, Althoff turned to one of his closest confidants, Bernhard Spinola, the administrative director of Berlin's *Charité* hospital. Spinola met with Heubner and was quite impressed by the Saxon candidate: "Last Thursday, Professor Heubner was with me for quite a long time, I really liked him, and he seems to be willing to accept the position. I have promised him the best possible consideration of his wishes from the side of the *Charité*."[63] Heubner's appointment followed soon after Spinola's positive assessment: four months after their meeting, on 30 April 1894, he gave his inaugural lecture in Berlin.[64]

The Heubner case shows two things. In the first place, it draws attention to the limits of the influence of individual letters of recommendation. If Althoff received unanimous advice, he would often follow it. When the advice was mixed, however, he had to disregard some of the recommendations and trust either his own judgement or that of his very closest confidants, like Spinola. Secondly, the Heubner case shows a lack of agreement on what qualities required loyal support and what qualities called for critical distance. Külz's appreciation of Heubner's publication record and managerial skills did not translate into an altogether positive assessment. In his eyes, his praiseworthy qualities were offset by his provincial appearance. Krehl recognized Heubner's typically Saxon demeanour as well but did not put as much weight on an upper-class, metropolitan self-presentation as Külz. Heubner's publication record and experience as a director of two clinics were much more relevant to Krehl's assessment. In the end, however, Althoff was free to make the final decision as he pleased.

Althoff's Full Force

The section on the Medical Faculty in Breslau illustrates the smoothest possible collaboration between Althoff and unanimous confidants. The following section on Heubner's appointment shows the complexity of the decision-making process when his confidants did not agree. In this section, I look at the role of letters of recommendation in an even more complex appointment procedure: one in which Althoff's aims are at odds with both the advice of his confidants and the desires of the faculty in question. This contested procedure concerns the eventual appointment of Emil Behring as professor of Hygiene at the University of Marburg. His appointment was both advocated and challenged because of his unusual background and the fact that he seemed to have great strengths as well as very serious shortcomings.

After the success of his diphtheria serum, Behring felt increasingly dissatisfied with his subordinate positions at Koch's institute.[65] In the winter of 1894, he accepted a temporary professorship in Halle.[66] Because his relationship with Koch deteriorated further during his time in Halle, Behring informed Althoff that he did not want to return to working at the Institute for Infectious Diseases.[67] He had his mind set on an appointment as professor of Hygiene at a Prussian university and threatened to leave the Prussian university system if his demand was not met. He drew attention to the attractive offers he had already received from other parties: he had been invited to direct a soon to-be-established serum institute in Budapest, was asked to become director of research at *Höchster Farbwerke*, the firm that mass-produced his diphtheria serum, and had been offered the opportunity to continue his research in Petersburg.[68] His preference, however, was a professorship in Marburg—the Prussian university city closest to Höchst.[69] There were, however, no vacancies in Marburg. Because Althoff wanted to preserve Behring for the Prussian university system without being able to offer him his desired professorship on short notice, he advised him to go to Höchst, so that he would at least remain in Germany.[70]

Because there was no vacancy for a Professor of Hygiene in Marburg, Althoff explored the possibility to create a special "serum professorship" for Behring at a Prussian university.[71] His confidant at the Faculty of Medicine in Bonn, the pharmacologist Karl Binz, was willing to champion the idea at his university.[72] But before Binz could accomplish anything in Bonn, new opportunities emerged in Marburg. The professor for Hygiene, Carl Fraenkel, accepted a professorship in Halle and his position became vacant. Even though this offered a great opportunity to reward Behring's serum research with his desired professorship, Fraenkel's transfer to Halle also illustrated Behring's shortcomings. After all, he had spent a full semester teaching in Halle but none of his colleagues had expressed any interest in retaining him for their university. He had failed to convince any of his peers of his merits as a university teacher.

This failure was somewhat surprising because before Behring's short tenure in Halle, Althoff had only received favourable assessments of his teaching. As a military doctor—who had never written a *Habilitation*—his early teaching experience was limited to courses for other military doctors.[73] Hermann Schaper, the medical director of the *Charité*, testified that Behring always "took great pains with his course, so that the chief staff doctors and the staff doctors have always attended it with the greatest interest."[74] A former student remembered the bacteriological courses that Behring had taught with the hygienist Bernhard Nocht. Nocht would read the lecture, after which both men would supervise the practical component of the course. He recalled that "Behring was very detailed, precise, and clear in his instruction" and added that he "had a great ability to explain something to the students, some of whom were completely inexperienced."[75] Heinrich Bonhoff, a former colleague at the Institute for Infectious Diseases, was even more enthusiastic: "[His] lecture was steady, calm, and strictly factual, easy to understand for everyone, with exact consideration of the understanding of the audience. [...] I have obtained the conviction that [he] has outstanding teaching skills."[76]

Soon, however, Althoff started to receive unfavourable assessments of Behring's teaching as well. The neurologist Eduard Hitzig, who had earlier recommended Behring for the Halle position, reported that two different sources had informed him that Behring was "somewhat meshuga."[77] His Halle colleague Josef von Mering criticized the narrowness of his teaching: "Professor Behring is an outstanding bacteriologist, but it is very doubtful, whether he is knowledgeable about the other fields of hygiene. As a teacher, he can at best claim modest success, which might partly be because of the fact that he has not lectured before, because he has not been habilitated, and partly to the fact that he does not yet master the topic completely."[78] In his eyes, Behring was only fit to supervise practical training. The pharmacologist Erich Harnack added that Behring "will only be a successful teacher for those who specifically work according to his intentions under his leadership."[79] These reports also reached Marburg, where they were seen as "extraordinarily unfavourable."[80]

Another reason for the Marburg Faculty of Medicine to be suspicious of Behring was the polemic disposition that he had displayed in a dispute with Virchow. Virchow had initially been critical of the diphtheria serum but eventually decided to introduce it to the Berlin children's hospital of which he was a co-founder, the *Kaiser- und Kaiserin-Friedrich-Kinderkrankenhaus*.[81] When a local newspaper reported that Virchow had claimed that the first successful experiments with the serum had been carried out in his hospital by one of his students, Behring was forced to defend his priority claim in public as well.[82] In a vitriolic article he called Virchow a "medical doctrinaire" who had to be opposed, characterized his understanding of the origins and cures of diseases as "heresies," and argued that his "dogmatism" had resulted in an "inquisition."[83] Even if Behring's medical arguments were quite convincing, his name calling was resented. He even admitted as much to Althoff.[84] Fraenkel informed Althoff that it was indeed this specific publication that had harmed Behring's chances in Marburg.[85]

In this light, it was hardly surprising that Behring was not shortlisted in Marburg. The faculty's list consisted of Albrecht Kossel and the former Koch associates Erwin von Esmarch and Moritz Eduard Cramer.[86] Althoff had instructed Fraenkel to make sure that Behring would be added to the list, but this proved to be unfeasible: "To get Behring on the list was simply impossible; except for me, nobody stood up for him."[87] Because Fraenkel knew that Althoff was also working on a special serum professorship for Behring in Bonn, he felt that it was not necessary to go against the wishes of his whole faculty.[88] Kossel was appointed and the unexpected opportunity to secure a Marburg professorship for Behring seemed to have passed.

Only one month later, however, a new opportunity occurred. In January 1895, Eduard Külz, Althoff's confidant and Marburg Professor of Physiology, died. Kossel took over Külz's chair and the search for a new professor for Hygiene started all over again.[89] Behring reminded Althoff of the overtures from Budapest and Petersburg and repeated his preference for a Marburg professorship.[90] Althoff, again, instructed Fraenkel to make sure that Behring would be included in the faculty's list of candidates. Fraenkel, again, could not deliver: "The candidacy of [Behring] seemed pointless to me. [...] A sudden [...] resort to Behring was really hardly possible for me, because I could not make any argument for it at all."[91] The faculty's list was the same as the one it had drawn up one month earlier, except for the inclusion of Walther Kruse instead of Kossel.[92]

Before the negotiations had reached this new stalemate, Behring had left for a long vacation in France. Althoff used his personal clout over the minister to delay the decision-making process and advocate for the appointment of Behring against the explicit wishes of the Marburg faculty. When Behring returned from his trip, the bad news was sent to Marburg: "[Y]our suggestions could not be considered, mainly because it was particularly important to retain a distinguished man like Professor Behring in the service of a Prussian university and there was no other option available at the time."[93] The medical faculty had lost its battle against the appointment of a man they considered to be unfit and Althoff had succeeded in keeping the inventor of the diphtheria serum in Prussia. It soon became clear that the faculty's suspicions about Behring's teaching were justified but this problem was soon resolved: Behring convinced Althoff to appoint his trusted Berlin colleague Erich Wernicke to take over most of his teaching.[94]

The case of Behring sheds further light on the professional and personal qualities that encouraged scholars to either loyally support their colleagues or to keep them at a distance. What is most striking is the fact that Behring's undeniable merits as a researcher did not carry a particularly heavy weight in the considerations of the Marburg Faculty of Medicine. Their assessment of the suitability of potential colleagues was mostly based on his teaching skills and personal character instead. Behring's qualities as a researcher did, however, figure prominently in Althoff's conviction that talented men like Behring should be retained for the Prussian university system, even if this would damage his personal relationship with the representatives of hesitant or even hostile faculties. Behring's troublesome appointment therefore underlines the way in which different focuses of loyalty could come into conflict with another. Althoff and some of Behring's associates may have felt a personal loyalty towards him. The representatives of the University of Marburg, however, were primarily guided by an institutional loyalty towards their own faculty. And even if Althoff must have felt some loyalty towards Behring, his primary loyalty was not to any individual scientist, faculty, or university, but to the Prussian university system as a whole. The contradicting recommendations of Althoff's informants and faculty members are best understood in the light of the existence of such different focuses of academic loyalty.

Professorship and Moral Responsibility

In the preceding chapter, I have drawn attention to Theodor Nöldeke's and Wilhelm Wundt's strikingly harsh reviews of books that discussed theological and philosophical topics.[95] Both men criticized what they perceived to be dogmatic speculation. One reason to be especially critical of theological and philosophical treatises is that these works might not only represent new developments in scholarship but also touch on issues of a very general interest such as questions of everyday morality and the appropriate role of the church and religion in society. The potentially far-reaching impact of theological and philosophical publications was recognized by critical reviewers like Nöldeke and Wundt. At the Berlin Ministry of Education, special attention was given to the moral character of those considered for a theological or philosophical professorship.

The special attention devoted to philosopher's religious and moral qualities becomes particularly clear when we compare two extensive disciplinary overviews in Althoff's papers: a report on German psychiatrists and neurologists by the Berlin *Privatdozent* Ernst Siemerling from 1889 and an overview of German philosophers by the Halle neo-Kantian Hans Vaihinger from 1893. Siemerling records the marital status and religious affiliation of the twenty-six scholars on his list, but he does not refer to these things in his assessments. His most striking observations are critical insights into the pettier elements of the personality of his peers. Otto Binswanger, for example, is described as follows: "Thinks very highly of himself; talks a lot, his statements are not very trustworthy, he is very secretive and always thinks of his own interest. Not a candid character." Eduard Hitzig receives a similar harsh treatment: "Has a very brusque, unsociable character, an egoist and autocrat through and through." The Breslau psychiatrist Carl Wernicke had "made himself unpopular because of his brusque behaviour" and was said to have a "dubious character." By far the biggest part of his overview, however, is taken up by lists of everyone's main publications, sometimes with short assessments such as "nothing new," "not very important," "good work with new insights," or simply "good."[96]

Religion is much more central to Vaihinger's assessment of German philosophers.[97] His overview suggests that their particular religious affiliation is not in itself important. What matters, though, is the way in which religion shapes their outlooks on life. He divides his colleagues into four groups on the basis of their attitudes towards religion: those who "have a positive religious interest," those who "support religious liberalism," those who "are radical in religious questions, all the way up to hostility against religion," and those who are "religiously indifferent." According to Vaihinger's classification, these different attitudes correlate with specific philosophical schools of thought. He notes that adherents to "apriorism" (*Apriorismus*), such as Kantians and Hegelians, tend to have a "positive religious interest" even if a few members of this school are labelled "liberal" or even "radical." Empiricists, he continues, tend to be liberal, radical, or indifferent. Wundt, who was labelled as a proponent of "critical empiricism," is categorized as either liberal or indifferent. His associate Richard Avenarius is described as a representative of the school of "positivist empiricism" and a religious radical.[98]

His radical reputation was probably the reason why he was never offered a professorship at a Prussian university. His Prussian colleagues were well-aware of his existence. In 1884, he was shortlisted by the University of Kiel to succeed Bruno Erdmann, who had moved to Breslau. As usual, Althoff carefully collected information about all candidates. In addition to Avenarius, these were the Halle *Extraordinarius* Gustav Glogau, the Bonn *Privatdozent* Theodor Lipps, and Hans Vaihinger, who was *Extraordinarius* in Strasbourg at the time.[99] To be sure, Avenarius religious radicalism was not the only criticism directed at him. Two of Althoff's informants drew attention to his limited publication output. The Jena philosopher Otto Liebman emphasized that at forty years Avenarius had "only published two slim volumes."[100] His Strasbourg colleague Wilhelm Windelband argued that his founding of the *Vierteljahrsschrift für Wissenschaftliche Philosophie* had come at a high cost: "Because the journal requires a lot of work, he has, alas, not accomplished anything for years, but I consider him to be very able and industrious; something will come out of it eventually."[101]

Avenarius' perceived religious shortcomings were a major obstacle to a professional appointment as well. This becomes particularly clear in the assessments that Althoff collected of his rivals. The Strasbourg professor Theobald Ziegler, for example, was criticized by his Marburg colleague Julius Bergmann on the grounds that he would be an advocate of the work of David Friedrich Strauss. Strauss' studies of life of Jesus were widely considered to be a threat to the Christian faith when they first appeared in the 1830s.[102] In the 1880s, Bergmann still warned that "the representation of this school of thought at a university by a full professor must pose a great danger to the students."[103] Like Bergmann, the curator of the

University of Halle, Wilhelm Schrader, did not explicitly refer to Avenarius but emphasized the "ethical-religious" views of one of his rivals, Gustav Glogau, and approvingly noted that the sincerity of his piety was illustrated by the fact that he had witnessed him attend church.[104]

Althoff received only one wholehearted endorsement of Avenarius' candidacy and it is quite telling that it was presented as an argument against potential doubts about Avenarius' moral uprightness. The recommendation by the Berlin philosopher Friedrich Paulsen started from the assumption that Althoff might suspect "just about any indiscretions or provocations from Avenarius that can disrupt the peace in public education."[105] He tried to reassure him: "I am convinced that whatever he says, will be said with the earnestness and objectivity that should be expected from a lecture dealing with the final and highest things. I think he is as incapable of defamation as of hypocrisy." All other informants, however, distanced themselves from Avenarius. Windelband subtly emphasized that he himself—of course—belonged to a very different intellectual tradition.[106] The Jena philosopher Otto Liebmann simply noted that Avenarius represented "a very extreme school of thought."[107] Bergmann recounts that he had never finished reading Avenarius' work, because he immediately recognized it as a pointless, superficial, and empiricist take on Kant.

Paulsen's recommendation did not weigh the scales in Avenarius' favour. Appointing him would have been too controversial. After all, even his supporters distanced themselves from his ideas and his opponents did not have to go out of their way to underline his shortcomings: it was enough to mention religion, morality, or the extremeness of his thought. A short reference to his small scholarly output would strengthen the case against him even further. In the end, Glogau was appointed in Kiel, Vaihinger succeeded Glogau in Halle, and Avenarius would remain in Zürich for the remainder of his career. In August 1896, the University of Freiburg was the first and only German university to include him at the top of their list of professorial candidates.[108] Soon after receiving the news that he might finally be called back to Germany, however, Avenarius passed away.[109]

Schools of Thought

Avenarius seems to have been unaware of the specificity of his colleagues' complaints about his supposed radicalism and irreligious amorality. In a letter to Wundt, he blamed his inability to find a job in Germany on prejudices against what he called the "Wundtian" school of thought.[110] Even though it is far from obvious that Avenarius is best understood as a true *Wundtianer*, the idea that your association with a certain school of thought could shape your career prospects is reasonable.[111] Often, the choice for a specific professorial candidate was also a choice for a specific approach to research and teaching. When Max Rubner succeeded Robert Koch as director of the Institute for Infectious Diseases, this signalled a clear change from an emphasis on bacteriology to a focus on physiology.[112] The generation of orientalists that followed Nöldeke's strove to "open the doors to a wider, deeper and more powerful Orient" in response to what they saw as the overly narrow positivist approach of the older generation.[113] This inspired an increasing engagement with contemporary Asian societies, cultural practices, and living languages in the late nineteenth and early twentieth centuries.

The philosophical discipline was highly susceptible to disputes between the proponents of different schools of thought as well. Vaihinger's overview of German philosophers listed four main schools of aprioristic philosophy.[114] These were all built on the approaches and insights of a leading philosopher from the past: Hegel, Herbart, Lotze, and Kant. Vaihinger also distinguished four main schools of empiricism: critical empiricism, psychological empiricism, idealistic empiricism, and positivist empiricism. He also added a section on "ultramontane Catholic philosophers" as well as a group of miscellaneous young philosophers, whose views had not yet fully crystalized and who could therefore not be categorized yet. Vaihinger's overview therefore suggests the existence of at least ten competing schools of philosophy in late nineteenth-century Germany.

There is no reason to assume that all Vaihinger's contemporaries would have agreed with his categorization. After all, there are many ways in which philosophers can be distinguished from each other. His overview does, however, give an indication of the variety of thought among German philosophers. Perceptions of a shared or radically different philosophical tradition also shaped professorial appointment procedures. Considerations about a candidate's suitability for a professorial chair cannot always be clearly distinguished from declarations of loyalty towards representatives of one's own school of thought. The search for a new professor of Philosophy in Berlin in 1894 illustrates the importance of the candidates' familiarity with different intellectual traditions. Even if the Faculty of Philosophy was looking for a modern thinker with both experimental and psychological competences, the eventual choice for Carl Stumpf was primarily based on the way in which his approach differed from both Wundt's and Georg Elias Müller's and the way in which it fitted the interests and sensibilities of his new Berlin colleagues.

Even though Stumpf and Wundt shared an interest in experimental research into the physiology of perception, they were not close to each other. In the early 1890s, they got involved in a debate about Weber's law, which quickly turned into a heated exchange of personal insults. The initial cause of their argument was the dissertation of Wundt's student Carl Lorenz.[115] Stumpf published an extensive criticism of Lorenz's findings, which encouraged Wundt to write an elaborate response to defend the work that had been carried out under his supervision and in his laboratory.[116] He did not limit himself to defending Lorenz's work but also added some snide jabs at Stumpf. He ended his exposition with the observation that his remarks would benefit Stumpf because they would teach him "to value, not only as the best, but also as the most useful virtue for a scientific researcher, this: to be fair towards others, to be stern towards himself."[117] This opened the floodgates to further personal recriminations. In his reply, Stumpf accused Wundt of a "mixture of untrue assertions, of confusions, of mutilations of the course of my thought, of obscure imputations and negligences, of infirm evasions, of fallacies of every kind, and of frequent assurances of the incapacity and ignorance of his adversity."[118]

This confrontation with Wundt probably worked to Stumpf's advantage in Berlin. There were widespread rumours that the influential Berlin professor of Physics, Hermann von Helmholtz, had fired Wundt when he was his assistant in Heidelberg in the 1860s because of his lacklustre mathematical prowess. These rumours have been contradicted by both Wundt and recent scholars.[119] At least one letter among Althoff's papers, however, suggests that the relation between Helmholtz and Wundt was strained at best. The Halle university librarian Otto Hartwig reminded the *Ministerialdirektor* of the rumours and added that "because the greatest men in Berlin, Helmholtz in particular, have fallen out with Wundt, his former pupil from Heidelberg, Stumpf is already for that reason better liked by them."[120]

The dispute between Wundt and Stumpf was not simply an example of name calling but also points at different schools of thought among scholars whose research interests largely overlapped. Lorenz and Stumpf had both investigated the ability to judge the middle tone between two tones sounded in series.[121] Lorenz concluded that listeners would pick out the arithmetic mean instead of a harmonious musical interval. Stumpf, on the other hand, argued that a well-trained listener would choose the musical mean as the middle tone instead. The most likely explanation for their different findings is that Lorenz and Wundt relied on a large data set obtained from musically untrained subjects, while Stumpf based his findings on the perception of a smaller number of trained music listeners.[122]

These differences illustrate how Stumpf's approach to the physiology of perception explicitly allowed for an appreciation of aesthetic judgement. In the eyes of some of his advocates, it was exactly this aesthetic appreciation that distinguished his work from what they considered to be Wundt's reductive positivism. Stumpf emphasized his aesthetic interests in a letter to the influential Berlin philosopher Wilhelm Dilthey as well: "In regard to the lectures, I will try to read aesthetics alongside psychology and for both I hope to find a growing audience that is not motivated by worries about examinations."[123] Stumpf's aesthetic sensibilities were also underlined by Althoff's most active advisor on the Berlin vacancy, the Breslau philosopher Theodor Lipps. He recommended Stumpf wholeheartedly, not least because "[a]bove all, it appears important to me that, thanks to Stumpf, justice will also be done to aesthetics."[124]

The unwillingness to consider Wundt—or any of his closest associates—limited the options for a philosopher with both experimental and psychological expertise. The possible candidacy of Hermann Ebbinghaus, who had founded the Berlin psychological laboratory, was hardly considered because he was seen as too limited in his experimental orientation and had a difficult relationship with Dilthey.[125] Even if the relationship between the two men would have been less strained, Ebbinghaus would still have faced an uphill struggle. Dilthey had already tried to convince Althoff to appoint Stumpf in Berlin as early as 1884.[126] The only other serious candidate was the Göttingen philosopher Georg Elias Müller. However, his colleague Ulrich von Wilamowitz-Moellendorf described him as a man who would have been called an "inconvenient subordinate" in the military and an "academic porcupine" but he was widely recognized as one of Germany's most prominent experimental philosophers as well.[127] Wilamowitz also added that Müller had become more sociable after he had left behind poverty, overcome prolonged disease, finished a book, and finally married.

Wilamowitz's characterization of Müller as an "academic porcupine" may not have damaged his chances in Berlin because some of Althoff's informants were quite critical of Stumpf's personality as well. Hartwig noted that Stumpf was "a very arrogant gentleman and therefore rather irritable and morose."[128] The Halle Professor of Law Friedrich Schollmeyer had reservations too: "Personally, I have the impression of a very tense human being and that has been confirmed to me by people who have had him as examiner."[129] Stumpf's and Müller's teaching skills also received little praise. Hartwig noted that Stumpf's lectures were well-attended but not as good as those of his Halle predecessor Benno Erdmann.[130] The legal scholar and librarian Hans Paalzow observed that his lectures were "carefully prepared" and "rich in ideas" but also noted that most attendants of the first lecture did not return for its follow-up.[131] Müller struggled with similar problems. The Göttingen mathematician Felix Klein reported that his lectures were well-attended but also emphasized that he seemed to be unable to convince any students to write their dissertation under his supervision.[132]

Both men were lavishly praised, however, for their accomplishments as independent thinkers and skilled experimentalists. Even though Stumpf had been a student of Franz Brentano, whose insights were hardly widely accepted in the academic mainstream, one report emphasized that Stumpf and Müller clearly belonged to the same intellectual tradition.[133] Wilhelm Schrader suggested that a consideration of their respective schools of thought would not be useful because both should be classified as belonging to "that side of Lotze's school that follows the so-called exact psychological investigation [...] without forsaking ideality."[134] Some of Althoff's other informants, however, disagreed and argued that Müller was hardly a true philosopher but rather a physiologist with only a limited interest in philosophy. Klein stated that Müller's true goal was "the creation of a *psychophysics* that corresponds with all our knowledge of today."[135] Lipps argued that "Müller's actual literary endeavours [...] throughout many years concerned *physiology*. [...] It is indeed to be feared [...] that physiology will eventually completely engross him. In any case, for Müller, *philosophy* is not at the *centre* of his interest."[136] Wilamowitz added that Müller "places himself in the natural sciences, and I believe that he trains the students who affiliate themselves especially with him somewhat one-sidedly."[137]

Stumpf was perceived to be less one-sided. Lipps stated that Stumpf had the lofty ambition to come to "a comprehensive intellectually and ethically satisfying perception of the world and of life."[138] Max Dessoir described Stumpf as a philosopher with a "brilliant general intuition" who uses "essentially logical tools to approach the questions of psychic life."[139] Dilthey emphasized Stumpf's attempts to "harmonize the spirit of the natural sciences with the highest interest of humankind" and enthusiastically added that "in this deeper relationship with religion and ethics [Stumpf] is unique among the scientific philosophers."[140] In the light of these observations, Schrader's assessment that Stumpf and Müller represented the same school seems to be incorrect. Müller's interest in experimental psychology was the result of a primary interest in psychophysics and physiology, while Stumpf's experimental interests were intimately connected with aesthetic, religious, and ethical questions. Because most of the members of the Berlin Faculty of Philosophy valued an interest in such questions over experimental prowess—and because they also trusted in Stumpf's ability to teach not only psychology but also history of philosophy, they showed a clear preference for Stumpf, who would eventually get the job.[141]

Loyalty and Independence in the Althoff System

The case studies in this chapter show that the system of informants and letters of recommendation created by Althoff served as another arena in which independence and—especially—loyalty could be expressed. It is important to note, however, that the letters of recommendation consulted for this chapter are quite different from modern-day letters of recommendation. Today, such letters are usually written at the request of a job applicant, but the letters discussed in this chapter were all written at the request of the man who made the hiring decisions. Therefore, these letters are not just favourable reflections on the merits of a single candidate, but rather assessments of several candidates as well as the perceived needs of the hiring institution and occasionally even the needs of the Prussian university system as a whole. The juxtaposition of the interests of different individuals and institutions created a discursive space within which loyalty and independence could be expressed. These expressions of loyalty and independence were particularly meaningful because they could decisively shape individual careers.

The different foci of loyalty in the letters of recommendation sent to Althoff complement the observations about the different audiences of book reviews in the preceding chapter. In that chapter, I argue that reviews are simultaneously directed at the reviewee and at the peer group as a whole. Therefore, the book review is a medium that is eminently suitable to create and maintain a group identity by showing loyal enthusiasm for certain authors and keeping an independent, critical distance from others.[142] Althoff's correspondence adds another piece to the complex puzzle of loyalties among scholars. In some respects, institutional loyalty towards a university, faculty, or research institute might be more important than loyalty towards individuals or the whole peer group. Fritsch's recommendation to refrain from appointing a successor for Friedrich Müller is a clear example of institutional loyalty outweighing personal loyalties. The unwillingness of all members of the Marburg Faculty of Medicine to appoint Behring is an example of institutional loyalty outweighing the national interest to find a fitting position of one of Prussia's most successful medical researchers.

This last example points at a surprising finding: assessments of research output and scholarly publications were not particularly important in most reflections on professorial candidates. The most plausible explanation for this relative neglect of research is the fact that most of Althoff's informants tried to find a balance between the specific needs of the hiring institutions and the qualities of the candidates. Research output was not seen as a quality that was particularly relevant to an individual scholar's functioning within an institutional context. Qualities that were considered relevant were teaching skills (e.g. Behring), financial situation (e.g. Küstner), social profile (e.g. Heubner), school of thought (e.g. Stumpf), and religious inclination (e.g. Avenarius). To some extent, the weight of different qualities differed between disciplines; a medical professor had to be skilled in managing a clinic and maintaining good relations with patients, while a philosopher's career opportunities were more dependent on his school of thought and his attitude towards religion.

This close look at the correspondence between Althoff and a wide circle of German academics does not only draw attention to the painstaking ways in which different kind of loyalties had to be balanced. It also points at the way in which the decisive involvement of third parties could provide individuals with a certain measure of independence from the power structures of the faculties and universities. One of these third parties is—obviously—the government. If it was not for Althoff's forceful intervention, Behring might not have been appointed to a Prussian professorial chair. His correspondence with Althoff highlights the potential influence of another third party as well: the industry. The career opportunities offered by the *Höchster Farbwerke* proved to be helpful in pressuring Althoff to enforce his appointment in Marburg. In the next chapter, I will—among other things—take a closer look at the way in which non-academic institutions and agents shaped the way in which scholars were able to maintain a delicate balance between loyal collegiality and critical independence.[143]

Notes

1. See, Chap. 3, 64.
2. See, Chap. 4, 126–129.
3. A noteworthy exception is: Michèle Lamont, *How Professors Think: Inside the Curious World of Academic Judgment* (Cambridge, MA: Harvard University Press, 2009).
4. R.M. Macleod, "The Royal Society and the Government Grant: Notes on the Administration of Scientific Research," *The Historical Journal* 14, no. 2 (1971): 328.
5. Ibid., 324, 329.
6. Maurice Crosland and Antonio Gálvez, "The Emergence of Research Grants within the Prize System of the French Academy of Sciences, 1795–1914," *Social Studies of Science* 19, no. 1 (1989): 94.
7. Rudolf Vierhaus, "Die Organisation wissenschaftlicher Arbeit: Gelehrte Sozietäten und Akademien im 18. Jahrhundert," in *Die Königlich Preußische Akademie der Wissenschaften zu Berlin im Kaiserreich*, ed. Jürgen Kocka, Rainer Hohlfeld and Peter Th. Walther (Berlin: Akademie Verlag, 1999), 19.
8. Walter Rüegg, "Ortsbestimmung: Die Königlich Preußische Akademie der Wissenschaften und der Aufstieg der Universitäten in den ersten zwei Dritteln des 19. Jahrhunderts," in *Die Königlich Preußische Akademie der Wissenschaften zu Berlin im Kaiserreich*, ed. Jürgen Kocka, Rainer Hohlfeld and Peter Th. Walther (Berlin: Akademie Verlag, 1999), 35.
9. Adolf von Harnack, *Geschichte der Königlich Preußischen Akademie der Wissenschaften zu Berlin* (Berlin: Georg Stilke, 1901), 770–774.
10. Bernhard vom Brocke, "Friedrich Althoff: A Great Figure in Higher Education Policy in Germany," *Minerva* 29, no. 3 (1991): 280.
11. Sabine Schilfert, "Friedrich Althoff—ein preußischer Geheimrat von Format? Bemerkungen zu einem wissenschafts-historischen Kolloquium anläßlich seines 150. Geburtstag," *Zentralblatt für das Bibliothekswesen* 103 (1989): 546.
12. Ibid., 446; Brocke, "Friedrich Althoff," 289.
13. Brocke, "Friedrich Althoff," 278.
14. Aleksandra Pawliczek, "Kontinuität des informellen Konsens: Die Berufungspolitik der Universität Berlin und ihre Dozenten im Kaiserreich und in der Weimarer Republik," in *Kontinuitäten und Diskontinuitäten in der Wissenschaftsgeschichte des 20. Jahrhunderts*, ed. Rüdiger von Bruch, Uta Gerhard tand Aleksandra Pawliczek (Stuttgart: Franz Steiner, 2006), 79.
15. Brocke, "Friedrich Althoff," 278.
16. Schilfert, "Friedrich Althoff," 546.

17. William S. Peirce and Peter Krüger, "Entrepreneurship in Bureaucracy: The Case of Friedrich Althoff," *Journal of Economic Studies* 20, no. 4/5 (1993): 68.
18. Regine Zott, "Hochgeehrter Herr Geheimer Rath! … Friedrich Theodor Althoffs Beziehungen zur Chemie," *Chemie in unserer Zeit* 42, no. 5 (2008): 323.
19. Friedrich Paulsen, *Die deutschen Universitäten und das Universitätsstudium* (Berlin: A. Asher & Co., 1902), 101.
20. Jurgen G. Backhaus, "The University as an Economic Institution: The Political Economy of the Althoff System," *Journal of Economic Studies* 20, no. 4/5 (1993): 13.
21. David E. Rowe, *A Richer Picture of Mathematics: The Göttingen Tradition and Beyond* (Cham: Springer, 2018), 12.
22. Brocke, "Friedrich Althoff," 284. Dessoir occasionally served as one of Althoff's informants as well. See this chapter, 163.
23. Backhaus, "The University as an Economic Institution," 13.
24. Hartmut Leppin, "Eduard Nordens Berufung nach Greifswald: Handlungsspielräume im 'System Althoff'," *Philologus* 142, no. 1 (1998): 167–169.
25. Gustav von Goßler to Reichskanzler von Bismarck, 8 April 1885, Geheimes Staatsarchiv Preußischer Kulturbesitz (hereafter GStA PK), VI. HA, Nl Althoff, No. 156.
26. Reichskanzler von Bismarck to Gustav von Goßler, 11 April 1885, GStA PK, VI. HA, Nl. Althoff, No. 156.
27. Karl Foy, "Zur Persönlichkeit Carl Arendt's: Einige Erinnerungen," *Ostasiatische Studien. Mittheilungen des Seminars für Orientalische Sprachen zu Berlin* 1 (1902): 181.
28. Professor Arendt über die Zustände am Seminar für Orientalische Sprachen. 4. VII 1889, GStA PK, VI. HA, Nl. Althoff, No. 156.
29. "Es wird windig."
30. Martin Hartmann to 'Sehr geehrter Herr Kreisschulinspektor,' 26 September 1889, GStA PK, VI. HA, Nl. Althoff, No. 156.
31. Georg Humbert to Friedrich Althoff, 20 January 1890 and 16 February 1890, GStA PK, VI. HA, Nl. Althoff, No. 156.
32. Ernst Kühl to Friedrich Althoff, 16 October 1890, GStA PK, VI. HA, Nl. Althoff, No. 156.
33. Heinrich Fritsch to Friedrich Althoff, 2 March 1891, GStA PK, VI. HA, Nl. Althoff, No. 112.
34. Heinrich Fritsch to Friedrich Althoff, 22 June 1892, GStA PK, VI. HA, Nl. Althoff, No. 112.

35. [Rector und Senat of the University of Breslau], *Chronik der Königlichen Universität zu Breslau für das Jahr vom 1. April 1890 bis zum 31. März 1891* (Breslau, 1891), 44.
36. Heinrich Fritsch to Friedrich Althoff, 9 August 1893, GStA PK, VI. HA, Nl. Althoff, No. 112.
37. Heinrich Fritsch to Friedrich Althoff, 1 May 1892, GStA PK, VI. HA, Nl. Althoff, No. 112.
38. Heinrich Fritsch to Friedrich Althoff, 22 June 1892, GStA PK, VI. HA, Nl. Althoff, No. 112. Fritsch's emphasis.
39. Heinrich Fritsch to Friedrich Althoff, 25 December 1892, GStA PK, VI. HA, Nl. Althoff, No. 112.
40. Note captioned "von Fritsch erhalten. A 27/11 93. Zu Hasse," GStA PK, VI. HA, Nl. Althoff, No. 112.
41. Heinrich Fritsch to Friedrich Althoff, 25 December 1892, GStA PK, VI. HA, Nl. Althoff, No. 112. Fritsch's emphases.
42. Heinrich Fritsch to Friedrich Althoff, 22 June 1892, GStA PK, VI. HA, Nl. Althoff, No. 112.
43. Heinrich Fritsch to Friedrich Althoff, 10 May 1893, GStA PK, VI. HA, Nl. Althoff, No. 112.
44. Friedrich Althoff to Heinrich Fritsch, 4 August 1893, GStA PK, VI. HA, Nl. Althoff, No. 112.
45. Hermann Kuhnt to Friedrich Althoff, 10 July 1893 and Bernhard Schultze to Friedrich Althoff, 11 July 1893, GStA PK, VI. HA, Nl. Althoff, No. 112.
46. See Chap. 2, 37–39.
47. Brocke, "Friedrich Althoff," 283.
48. Heinrich Fritsch to Friedrich Althoff, 9 August 1893, GStA PK, VI. HA, Nl. Althoff, No. 112.
49. Hugo Falkenheim to Friedrich Althoff, received on 14 November 1893, GStA PK, VI. HA, Nl. Althoff, No. 112.
50. Ludolf Krehl to Friedrich Althoff, 13 November 1893, GStA PK, VI. HA, Nl. Althoff, No. 112.
51. Eduard Külz to Friedrich Althoff, 12 November 1893, GStA PK, VI. HA, Nl. Althoff, No. 112. "Dresden" refers to the Saxon government that was located in this city.
52. Eduard Külz to Friedrich Althoff, 3 December 1893, GStA PK, VI. HA, Nl. Althoff, No. 112.
53. Ludolf Krehl to Friedrich Althoff, 13 November 1893, GStA PK, VI. HA, Nl. Althoff, No. 112.
54. Volker Hesse, "Leben und Werk Otto Heubners: Eine Würdigung des ersten Lehrstuhlinhabers der Kinderheilkunde in Deutschland anlässlich seines 90. Todestages," *Monatsschrift Kinderheilkunde* 12 (2016): 116.

55. Heinrich Fritsch to Friedrich Althoff, 9 August 1893, GStA PK, VI. HA, Nl. Althoff, No. 112.
56. Pawliczek, "Kontinuität des informellen Konsens," 86. This also explains Althoff's question about Soltmann's Jewishness.
57. Eduard Külz to Friedrich Althoff, 13 November 1893, GStA PK, VI. HA, Nl. Althoff, No. 112.
58. Eduard Külz to Friedrich Althoff, 3 December 1893, GStA PK, VI. HA, Nl. Althoff, No. 112.
59. Hugo Falkenheim to Friedrich Althoff, received on 14 November 1893, GStA PK, VI. HA, Nl. Althoff, No. 112.
60. Eduard Külz to Friedrich Althoff, 3 December 1893, GStA PK, VI. HA, Nl. Althoff, No. 112.
61. Hesse, "Leben und Werk Otto Heubners," 1116; Klaudia Tomasevic, Die medizinische Versorgung von Kindern Mitte des 19. Jahrhunderts am Beispiel von Würzburg (PhD diss., Bayererischen Julius-Maximilians-Universität zu Würzburg, 2002), 87.
62. Pawliczek, "Kontinuität des informellen Konsens," 87.
63. Bernhard Spinola to Friedrich Althoff, 26 December 1893, GStA PK, VI. HA, Nl. Althoff, No. 112.
64. Hesse, "Leben und Werk Otto Heubners," 1119.
65. The development of the diphtheria serum is discussed at more length in Chap. 2, 35–39.
66. Derek S. Linton, *Emil von Behring: Infectious Disease, Immunology, Serum Therapy* (Philadelphia, PA: American Philosophical Society, 2005), 102, 192.
67. Emil Behring to Friedrich Althoff, 24 December 1894, GStA PK, VI. HA, Nl. Althoff, No. 325.
68. Emil Behring to Friedrich Althoff, 2 January 1895, GStA PK, VI. HA, Nl. Althoff, No. 325.
69. Emil Behring to Friedrich Althoff, 8 February 1895, GStA PK, VI. HA, Nl. Althoff, No. 325.
70. Emil Behring to Friedrich Althoff, 30 December 1895, GStA PK, VI. HA, Nl. Althoff, No. 326.
71. Emil Behring to Friedrich Althoff, 25 February 1895, GStA PK, VI. HA, Nl. Althoff, No. 326.
72. Carl Binz to Friedrich Althoff, 29 January 1895, GStA PK, VI. HA, Nl. Althoff, No. 326.
73. On the challenge of finding permanent academic employment without a *Habilitation*, see also Chap. 3, 70–72.
74. Hermann Schaper to Friedrich Althoff, 8 August 1894, GStA PK, VI. HA, Nl. Althoff, No. 326.

75. Anonymous testimony of a former student of Behring, GStA PK, VI. HA, Nl. Althoff, No. 326.
76. Heinrich Bonhoff to Friedrich Althoff, 25 August 1894, GStA PK, VI. HA, Nl. Althoff, No. 326.
77. Eduard Hitzig to Friedrich Althoff, 10 August 1894, GStA PK, VI. HA, Nl. Althoff, No. 326.
78. Josef von Mering to Friedrich Althoff, 10 December 1894, GStA PK, VI. HA, Nl. Althoff, No. 326.
79. Erich Harnack to Friedrich Althoff, 13 December 1894, GStA PK, VI. HA, Nl. Althoff, No. 326.
80. Carl Fraenkel to Friedrich Althoff, 18 December 1894, GStA PK, VI. HA, Nl. Althoff, No. 326.
81. Rudolf Virchow to Friedrich Althoff, 17 October 1894, GStA PK, VI. HA, Nl. Althoff, No. 326.
82. Linton, *Emil von Behring*, 177.
83. Emil Behring, "Das neue Diphtheriemittel," *Die Zukunft*, neunter Band (1894).
84. Emil Behring to Friedrich Althoff, 6 December 1894, GStA PK, VI. HA, Nl. Althoff, No. 325.
85. Carl Fraenkel to Friedrich Althoff, 18 December 1894, GStA PK, VI. HA, Nl. Althoff, No. 326.
86. Carl Fraenkel to Friedrich Althoff, 25 Feburary 1895, GStA PK, VI. HA, Nl. Althoff, No. 326.
87. Carl Fraenkel to Friedrich Althoff, 18 December 1894, GStA PK, VI. HA, Nl. Althoff, No. 326.
88. Friedrich Althoff to Emil Behring, 30 December 1894, GStA PK, VI. HA, Nl. Althoff, No. 326.
89. Friedrich Althoff to Emil Behring, 31 January 1895, GStA PK, VI. HA, Nl. Althoff, No. 326.
90. Emil Behring to Friedrich Althoff, 8 February 1895, GStA PK, VI. HA, Nl. Althoff, No. 325.
91. Carl Fraenkel to Friedrich Althoff, 9 February 1895, GStA PK, VI. HA, Nl. Althoff, No. 326.
92. Friedrich Althoff to Emil Behring, 25 February 1895, GStA PK, VI. HA, Nl. Althoff, No. 326.
93. Der Königliche Kurator der Universität Marburg to die Medizinische Fakultät Marburg, 11 April 1895, Behring Nachlass digital: EvB/L271/5.
94. Within a year of his arrival in Marburg Behring already informed Wernicke that he was trying to bring him to Marburg: Emil Behring to Erich Wernicke, 29 January 1896, Staatsbibliothek Berlin, Preußischer Kulturbesitz (hereafter Stabi), Nl. 156, Erich Wernicke. Wernicke arrived

in Marburg in April 1896: Emil Behring to Erich Wernicke, 2 April 1896, Stabi, Nl. 156.

95. See Chap. 4, 126–127.
96. Siemerling: Psychiater und Neurologen, 1.3.89, GStA PK, VI. HA, Nl. Althoff, No. 92.
97. Übersicht über die philosophischen Universitätsdocente Deutschlands (mit Einschluss Österreichs u. d. Schweiz) nach ihren Richtungen (nebst einigen Notizen über deren Hauptwerke, Geburtsjahr, Heimat, Konfession u.s.w.) by Hans Vaihinger, October 1893, GStA PK, VI. HA, Nl. Althoff, No. 118.
98. For more about the relation and collaboration between Wundt and Richard Avenarius, see Chap. 3, 65–67.
99. Bruno Erdmann to Friedrich Althoff, 19 January 1884, GStA PK, VI. HA, Nl. Althoff, No. 118.
100. Otto Liebmann to Friedrich Althoff, 23 January 1884, GStA PK, VI. HA, Nl. Althoff, No. 118.
101. Wilhelm Windelband to Friedrich Althoff, 24 January 1884, GStA PK, VI. HA, Nl. Althoff, No. 118.
102. Erik Linstrum, "Strauss's Life of Jesus: Publication and the Politics of the German Public Sphere," *Journal of the History of Ideas* 71, no. 4 (2010): 605–606.
103. Julius Bergmann to Friedrich Althoff, 25 January 1884, GStA PK, VI. HA, Nl. Althoff, No. 118.
104. Wilhelm Schrader to Friedrich Althoff, 1 February 1884, GStA PK, VI. HA, Nl. Althoff, No. 118.
105. Friedrich Paulsent to Friedrich Althofff, 22 January 1884, GStA PK, VI. HA, Nl. Althoff, No. 118.
106. Wilhelm Windelband to Friedrich Althoff, 24 January 1884, GStA PK, VI. HA, Nl. Althoff, No. 118.
107. Otto Liebmann to Friedrich Althoff, 23 January 1884, GStA PK, VI. HA, Nl. Althoff, No. 118.
108. Hugo Münsterberg to Wilhelm Wundt, 12 August 1896, Universitätarchiv Leipzig (hereafter UAL): NA Wundt/III/701–800/767a/549–556.
109. Hugo Münsterberg to Wilhelm Wundt, 21 August 1896, UAL: NA Wundt/III/701–800/767b/557–572.
110. Richard Avenarius to Wilhelm Wundt, 2 February 1883, UAL: NA Wundt/III/1001–1100/1023/85–88.
111. On Avenarius' relation to the Wundtian tradition, see: Chiara Russo Kraus, "Back to the origins of the repudiation of Wundt: Oswald Külpe and Richard Avenarius," *Journal of the History of the Behavioral Sciences* 53, no. 1 (2017): 30–32.

112. Stephan Rössner, "Max Rubner (1854–1932)," *Obesity Review* 14 (2013): 432.
113. Suzanne L. Marchand, *German Orientalism in the Age of Empire: Religion, Race, and Scholarship* (Cambridge: Cambridge University Press, 2009), 212–216.
114. Übersicht über die philosophischen Universitätsdocente Deutschlands (mit Einschluss Österreichs u. d. Schweiz) nach ihren Richtungen (nebst einigen Notizen über deren Hauptwerke, Geburtsjahr, Heimat, Konfession u.s.w.) by Hans Vaihinger, October 1893, GStA PK, VI. HA, Nl. Althoff, No. 118.
115. Carl Lorenz, "Untersuchungen über die Auffassung von Tondistanzen," *Philosophische Studien* 6 (1890): 26–103.
116. Carl Stumpf, "C. Lorenz: Untersuchungen über die Auffassung von Tondistanzen," *Zeitschrift für Psychologie und Physiologie der Sinnesorgane* 1 (1890): 26–103; Wilhelm Wundt, "Ueber Vergleichungen von Tondistanzen," *Philosophische Studien* 6 (1891): 605–640.
117. Wundt, "Ueber Vergleichungen von Tondistanzen,"640.
118. Carl Stumpf, "Mein Schlußwort gegen Wundt," *Zeitschrift für Psychologie und Physiologie der Sinnesorgane* 2 (1891): 438–440. The English translation has been taken from: Edwin G. Boring, "The Psychology of Controversy," *The Psychological Review* 36, no. 2 (1929): 110.
119. The story about Wundt being fired by Helmholtz is most famously recounted in: G. Stanley Hall, *Founders of modern psychology* (New York, NY: Appleton and Company, 1912), 311. Wundt denies his account in: Wilhelm Wundt, *Erlebtes und Erkanntes* (Stuttgart: Alfred Kröner, 1920), 154. His criticism of Hall's account is corroborated in: Saulo de Freitas Araujo, "Wilhelm Wundt als Assistant von Hermann v. Helmholtz an der Medizinische Fakultät der Universität Heidelberg: Berichtigende Bemerkungen," in *Medizin-, Pharmazie- und Wissenschaftsgeschichte vom Mittelalter bis zur Gegenwart*, ed. Regine Pfrepper (Aachen: Shaker Verlag, 2007).
120. Otto Hartwig to Friedrich Althoff, 27 July 1893, GStA PK, VI. HA, Nl. Althoff, No. 118.
121. Alexandra E. Hui, "The bias of "music-infected consciousness": the aesthetics of listening in the laboratory and on the city streets of fin-de siècle Berlin and Vienna," *Journal of the History of the Behavioral Sciences* 48, no. 3 (2012): 241.
122. In Chap. 2, 47, I refer to subjects in Wundt's laboratory as well-trained. However, being trained in introspection is not the same as being trained in interpreting sound within the framework of music theory.
123. Carl Stumpf to Wilhelm Dilthey, 5 October 1893, GStA PK, VI. HA, Nl. Althoff, No. 118.

124. Theodor Lipps to Friedrich Althoff, 31 August 1893, GStA PK, VI. HA, Nl. Althoff, No. 118.
125. Lothar Spring, "The Berlin Psychological Tradition: Between Experiment and Quasi-Experimental Design, 1850–1990," in *World Views and Scientific Discipline Formation: Science Studies in the German Democratic Republic*, ed. William R. Woodward and Robert S. Cohen (Dordrecht: Springer Science + Business, 1991), 108–109.
126. Mitchell G. Ash, *Gestalt Psychology in German Culture, 1890–1967: Holism and the Quest for Objectivity* (Cambridge: Cambridge University Press, 1995), 31.
127. Ulrich von WIlamowitz-Moellendorf to Friedrich Althoff, 2 September 1893, GStA PK, VI. HA, Nl. Althoff, No. 118.
128. Otto Hartwig to Friedrich Althoff, 27 July 1893, GStA PK, VI. HA, Nl. Althoff, No. 118.
129. Friedrich Schollmeyer to Friedrich Althoff, 28 July 1893, GStA PK, VI. HA, Nl. Althoff, No. 118.
130. Otto Hartwig to Friedrich Althoff, 27 July 1893, GStA PK, VI. HA, Nl. Althoff, No. 118.
131. Hans Paalzow to Friedrich Althoff, 31 July 1893, GStA PK, VI. HA, Nl. Althoff, No. 118.
132. Felix Kleinto Friedrich Althoff, 30 July 1893, GStA PK, VI. HA, Nl. Althoff, No. 118.
133. Ash, *Gestalt Psychology*, 28–29.
134. Wilhelm Schrader to Friedrich Althoff, 28 July 1893, GStA PK, VI. HA, Nl. Althoff, No. 118.
135. Felix Klein to Friedrich Althoff, 30 July 1893, GStA PK, VI. HA, Nl. Althoff, No. 118. Klein's emphasis.
136. Theodor Lipps to Friedrich Althoff, 31 August 1893, GStA PK, VI. HA, Nl. Althoff, No. 118. Lipps's emphases.
137. Ulrich von WIlamowitz-Moellendorf to Friedrich Althoff, 2 September 1893, GStA PK, VI. HA, Nl. Althoff, No. 118.
138. Theodor Lipps to Friedrich Althoff, 31 August 1893, GStA PK, VI. HA, Nl. Althoff, No. 118.
139. Max Dessoir to Friedrich Althoff, no date, GStA PK, VI. HA, Nl. Althoff, No. 118.
140. Wilhelm Dilthey to Friedrich Althoff, no date, GStA PK, VI. HA, Nl. Althoff, No. 118.
141. Ash, *Gestalt Psychology*, 31.
142. See Chap. 4, 128–129.
143. See Chap. 6, 191–195.

CHAPTER 6

A Moral Economy of Scholarship: The Balance Between Loyalty and Independence

Moral Economies

In the preceding chapters, I investigate how considerations of loyalty and independence are central to a wide range of everyday practices of scholarship. In Chap. 2, I draw on private correspondences to outline the ways in which scholars provided critical support of each other's work in progress. Chapter 3 provides a look behind the scenes of three scholarly journals and illustrates the extent to which editorial decisions were shaped by the interests and expectations of such different stakeholders as publishers, contributors, and audiences. In Chap. 4, I offer a more substantive assessment of the way in which published book reviews allowed for expressions of loyalty and assertions of independence. Chapter 5, finally, examines how the Althoff system provided an environment that guaranteed continuous tensions between loyalties to individual scholars, academic institutions, and intellectual traditions. In this chapter, the emphasis will not be on additional detailed case studies of displays of loyalty or independence, but rather on highlighting the moral economy of scholarly evaluation, the outlines of which have become visible in the preceding chapters.

In the first chapter, I have reflected on the way in which the idea of a moral economy of scholarship can serve as a theoretical framework that allows for a better understanding of the way in which different scholarly virtues can be balanced.[1] I have drawn particular attention to Robert

© The Author(s), under exclusive license to Springer Nature Switzerland AG 2022

C. Engberts, *Scholarly Virtues in Nineteenth-Century Sciences and Humanities*, https://doi.org/10.1007/978-3-030-84566-7_6

Kohler's work on early twentieth-century fruit fly geneticists. For the case studies in this study, I was particularly inspired by Kohler's thick descriptions of scientific life that combine detailed biographies of individual *Drosophila* researchers, the everyday working arrangements in their laboratories, the personal relations between the most important researchers in the field and representatives of various other research centres, and the values and mutual expectations that emerged among the "fly people." Kohler's attention to individual, interpersonal, and institutional values and practices results in what Clifford Geertz might have recognized as a thick description in the sense of "a stratified hierarchy of meaningful structures."[2]

Even if loyalty and independence were not Kohler's primary interests, his study contains some pointed observations about the place of these virtues in early twentieth-century science. Kohler gives a detailed description of an "elite, bound by a shared sense of participating in a remarkable history" which shares "a moral ethos of cooperation and communality."[3] During the early years of fruit fly research, a clear sense of mutual loyalty existed among its practitioners.[4] However, as the community of *Drosophila* researchers grew, assertions of independence from an increasing number of "fly people" occurred and mutual trust eroded.[5] The peculiar moral economy characterized by strong feelings of loyal collegiality among a relatively small number of ambitious and innovative researchers only stayed in place for a few decades.

Patrick McCray's study about the sharing of large telescopes by late twentieth-century US astronomers shares Kohler's emphasis on detailed descriptions of everyday practices of scholarship.[6] The moral economy that he describes, however, is very different from that of the *Drosophila* community. While Kohler draws attention to the way in which an abundance of research materials and promising research questions encouraged the emergence of a tightly knit community of loyal researchers, McCray shows how a scarcity of resources encouraged competition and assertions of intellectual independence. The strict limitations on the availability of large telescopes—in combination with a fast-growing community of astronomers—resulted in a fragmented research community whose members were involved in a continuous competition for access to scarce but essential resources.

Kohler and McCray were not primarily interested in the way in which different values touched on each other. Their studies suggest, however, that mutual trust and an ethos of loyalty are more likely to thrive in small communities with plenty of resources than in larger communities with less resources. The observation that different environments encourage different attitudes towards values like loyalty and independence, then, suggests that the balance between these virtues is very likely to be subject to change. This need to find a balance between loyalty and independence is acknowledged in Lorraine Daston's conception of a moral economy as a "web of affect-saturated values" with "equilibrium points and constraints."[7] In the following sections, I will use the—Kohlerian—thick descriptions of practices of scholarship from the preceding chapters to give an overview of the many "equilibrium points and constraints" that characterized nineteenth-century German ideals and performances of scholarly loyalty and independence. As Kohler's and McCray's observations suggest, the tensions between both virtues were relatively easy to manage in the private sphere but posed more of a threat once a larger peer group, and a variety of academic and non-academic institutions got involved.

A superficial look at this chapter's structure would suggest that my analysis is based on a continuum model. The first section looks at close friends in academia; the second section investigates collaborative hierarchies; the third section looks at the broader peer group; the other sections explore the relations between individual scholars and a variety of academic and non-academic institutions. At first sight, this might give the impression that I am trying to fit all these people and institutions on one continuum that ranges from close intimate relationships to formal distant relations. This impression is, however, only justified to a limited extent. Of course, friends are closer to an individual scholar than the peer group as a whole and collaborators tend to be closer acquaintances than journal editors or state representatives. At the same time, however, it is important to acknowledge some rather obvious shortcomings of such a continuum model of closeness and distance.

The first shortcoming is that not everyone necessarily agrees on assessments of closeness and distance. Some scholars might feel particularly close to their peers with shared interests in faraway places, while other might identify more strongly with their colleagues at the institution at which they are employed. Some, like Wundt, might even feel most intimately related to friends from their student days who ended up working both in different fields of study and in different cities.[8] A second shortcoming of the continuum model is that it is in no way self-evident that all people and institutions discussed in this study fit on one single continuum of closeness and distance in any meaningful way. Everyday encounters with laboratory co-workers on the one hand and occasional exchanges with state representatives on the other hand are not necessarily more or less intimate variations of the same relationship. They can also be understood as two fundamentally different relations. While reading this chapter, it is therefore important to keep in mind that—even though its structure suggests a continuum model—its protagonists are very likely to disagree both about the placement of specific relationships on an imaginary continuum and about the usefulness of locating relations on such a continuum at all.

Friends as Colleagues

The first case study of the first chapter describes the relation between Theodor Nöldeke and Michael Jan de Goeje. This relation provides an interesting starting point of an analysis of loyalty and independence because only very few successful scholars shared such a warm personal and professional bond for such a long time as these two. "May this intimate harmony continue to exist for our whole lives and may we always be able to remain proud to be each other's heartfelt friends," De Goeje wrote in 1863.[9] Looking back on the early years of their friendship almost half a century later, Nöldeke gladly remembered the "immediate connection between us, even if we did not yet know how close and lasting our friendship would be."[10] If mutual loyalty can be expected to shape mutual evaluation among any two scholars discussed in this study, it is among these men. Unsurprisingly, their correspondence does indeed contain plenty of expressions of friendship, appreciation of each other's work, and mutual support. What is more interesting, however, in the light of my interest in striking a balance between loyalty and independence, is the way in which they dealt with topics about which they disagreed.

The long and exhaustive lists of corrections and suggested changes after reading each other's not yet published work present the most striking example of critical mutual evaluation. Nöldeke's critical proofreading of long excerpts of De Goeje's edition of al-Ṭabarī's *Annals* shows that he felt free to comment on every element of the manuscripts of his friend and his collaborators, whether it was grammar, metre, or the appropriate use of auxiliary sources. De Goeje gratefully accepted his detailed and well-argued suggestions, which undoubtedly contributed to the overall quality of the end product and its eventual favourable critical reception.

Nöldeke's and De Goeje's willingness to accept each other's views was not limited to such obviously useful critiques. In the early years of their friendship, for example, De Goeje expressed his disappointment that Nöldeke preferred studying a range of smaller Semitic languages to working on editions of classic Arabic texts that were not yet available in Europe: "You have worked with care and diligence and delivered what could be delivered, but has it been worth the hassle? Would you not have given the public much more important results if you would have dedicated that same amount of time and effort to Arabic things?"[11] Nöldeke casually dismissed his concerns: "You, dear boy, confuse your inclination with the importance of things. You probably do not have an inclination for geology or the history of Roman law; do you therefore think that these subjects are less important than yours? You surely do not! I ask for the same respect for my pet studies."[12] De Goeje immediately assured him that he was—of course—not that "narrow-minded" and never raised the issue again.[13]

The strength of their early friendship was soon put to the test again. The way in which they dealt with their disagreement about the merits of Reinhart Dozy's *De Israëliten te Mekka* (*The Israelites at Mecca*) illustrates the mutual tolerance that had developed between them. De Goeje praised his *Doktorvater*'s book, in which he presented daring speculations about the history of Mecca as a religious centre.[14] Nöldeke, however, was critical of both Dozy's book and De Goeje's appraisal: "It pains me that an intelligent man like Dozy delivers such unmethodological investigations; yet it pains me much more that you accept and praise without reservation, all this that I can only see as a sham."[15] Nöldeke and De Goeje would discuss Dozy's *Israëlieten* for more than a year. Both men stubbornly stuck to their opinion, but this did not damage the amicable character of their correspondence in any way.

The correspondence between Wilhelm Wundt and Gustav Theodor Fechner provides another example of the way in which a close personal relationship could encourage a critical assessment of each other's work. After Wundt's arrival at the University of Leipzig in 1875, the men developed a relation that was intimate enough for Fechner's widow to ask him to sort out her husband's papers after his death in 1887.[16] The mutual respect between Wundt and Fechner allowed them to critically distance themselves from elements of each other's work in private letters without damaging their amiable relationship. This is most strikingly illustrated by their discussion about the spiritist phenomena associated with Henry Slade, as discussed in the second chapter.[17]

In his first letter about this topic, Fechner emphasized that he preferred to discuss the issue in private rather than in public.[18] This privacy created the right setting for a candid debate. If Fechner would have chosen to take Wundt's scathing criticism of the trustworthiness of the witnesses of Slade's performance as the starting point of their discussion, their discussion might have been less amiable. After all, Wundt's harsh critique of the witnesses of Slade's apparent miracles was particularly pertinent to Fechner, who suffered from declining eyesight.[19] In his letters to Wundt, Fechner did not dwell on his own perceptiveness, however, but mostly drew attention to the trustworthiness of the other observers. He defended his peers and accused Wundt of spreading "suspicion of the investigations so far based on vague assumptions."[20] Even if Wundt stuck to his opinion about the utter untrustworthiness of Slade's claims, he maintained a "friendly, amiable" tone and both men happily agreed to disagree.[21] Their later discussions of the work carried out in Wundt's laboratory show a similar benevolent acceptance of each other's different opinions.[22]

The examples of Nöldeke, De Goeje, Wundt, and Fechner show that it was relatively easy for good friends to combine loyal support with independent criticism. It must be noted, however, that their continuous amiable relations can be attributed to certain particularly favourable conditions. First, their collaborations usually only consisted of commenting on each other's work and rarely amounted to working on the same project. Nöldeke's contributions to the al-Ṭabarī edition is the only exception—and this was only a minor part of his research output in the 1870s. Second, none of them were in direct competition for university appointments or funding. Nöldeke never considered a permanent move to the Netherlands and when De Goeje could not find a suitable university position in the Netherlands, he considered quitting academic life altogether but never suggested trying his luck in Germany.[23] Wundt and Fechner did not compete either because they were at different points in their careers: Fechner retired when Wundt arrived in Leipzig. The relation between Emil Behring and Erich Wernicke, however, proves that balancing loyalty and independence was more challenging between friends who were working at the same project at the same point in their career.

During the development of the diphtheria serum, Behring and Wernicke became close friends. In November 1891, Behring even suggested that they should move in together: "I promise you that I will not be on your back too much."[24] Three months later, he looked forward to even closer professional and personal rapprochement in the light of his assessment that they were "already semi-married, after all."[25] Wernicke also expressed his friendship in Behring's guest book in 1897 by adapting a line from a

Schiller poem about loyalty: "Friendship is not an empty illusion!"[26] A superficial look at their correspondence also suggests that their friendship withstood candid criticism. Behring repeatedly reminded Wernicke of the fact that he was making himself too dependent on him. If Wernicke really wanted to pursue an academic career, Behring argued, he would have to make sure that he published independently about research that he had initiated, planned, and executed himself.[27] Behring's criticism already shows, however, that his relationship with Wernicke was not as equal as the relations between Nöldeke and De Goeje, or between Wundt and Fechner.

Wernicke became increasingly frustrated with the inequal relationship between himself and Behring. Two months before his guest book entry, he wrote his friend a letter in which he accused him of being too dominant. Behring patronizingly replied that he preferred to receive "many, but less reproachful letters."[28] Several years later, the financial success of the diphtheria serum further soured their relationship. Wernicke had hoped to receive a larger share of the serum's revenue than Bering allowed him but did not dare to challenge Behring's claim to own the "full scientific and financial rights of discovery."[29] Their friendship was restored after Wernicke left Marburg to become the director of the *Hygienisches Institut* (Hygienic Institute) of the University of Posen in 1899. Behring observed that "a temporary separation often clears up the relationship in a favourable way."[30]

Eight years later, when the troubled Behring had temporarily retreated to the *Neuwittelsbach* sanatorium in Munich, he called Wernicke "the most faithful among the faithful" in a letter full of self-pity and nostalgia.[31] After Behring's death in 1917, Wernicke stayed in touch with his widow. After visiting her in 1924, he left another message in the family's guest book. This time he pointed out how this visit had "renewed [his] old feelings of friendship."[32]

Students and Subordinates

The relation between Behring and Wernicke combined elements of friendship with characteristics of the relation between a superior and a subordinate. Behring's awareness of this is illustrated by his claims to be the sole discoverer of the diphtheria serum and his insistence that Wernicke should carry out independent research. Wernicke's powerlessness in the face of Behring's claims shows that he recognized this inequality as well. The power differences between Behring and Wernicke were more typical of academic collegiality than the equal relationships between Nöldeke and De Goeje. Nineteenth-century German academia was very hierarchical. David Cahan emphasizes the "strict patriarchal manner" by which full professors exerted their authority over students, *Privatdozenten* and *Extraordinarien*.[33]

Alexander Busch also underlines these power differences: "[H]e who controlled the institutes, the means of production in the field of research, [...] also controlled therewith the research workers, their opportunities and their prospects."[34] Jeffrey Johnson provides a more detailed picture of these power structures by distinguishing academic hierarchies, which describe the power differences between students and full professors, institutional hierarchies, which are relevant to the different positions of the scholars working in research institutes, and disciplinary disciplines, which are grounded in individual reputation.[35] These power differences could be particularly striking between *Doktorvaters* and their former students.

The relation between Robert Koch and his associates during the cholera outbreaks in Hamburg and Altona in the early 1890s provides a mutually beneficial example of such power inequalities. When cholera broke out in Hamburg in August 1892, the Berlin authorities ordered Koch to examine the situation. He only stayed in the city for one day and sent his former assistant and Giessen Professor of Hygiene, Georg Gaffky, to Hamburg to supervise the fight against cholera on the spot.[36] The following winter another former assistant, Richard Pfeiffer, was sent to neighbouring Altona to fight a new outbreak.[37] Gaffky and Pfeiffer did not primarily see themselves as servants of the city or the Prussian state authorities but rather as representatives of the Kochian approach to disease control. Gaffky's expression of loyalty was particularly explicit: "I certainly know that you trust in me, that I will always aim to work in your spirit [...]. Because you are the soul of it all and we are your executive bodies."[38]

During their stay in Hamburg and Altona, Gaffky and Pfeiffer kept Koch updated about all their findings and actions. This enabled Koch to support and direct their efforts from Berlin. The importance of his support is most obvious in his handling of Gaffky's complaints about the ministry's unresponsiveness to his request to have two army doctors appointed as assistants.[39] Koch had access to prominent government officials and shared Gaffky's reports with them.[40] After reading Gaffky's complaint, Karl Köhler, the director of the Imperial Health Office, immediately sent the requested army doctors to Hamburg.[41] At the same time, Koch also benefited himself from the loyal collaboration of Gaffky and Pfeiffer. Their data allowed him to gain the upper hand in an ongoing debate about the aetiology of cholera with the Munich hygienist Max von Pettenkofer.[42] Koch's descriptions of the Hamburg and Altona epidemics were often almost literally lifted from the reports of Gaffky and Pfeiffer.[43] In this way, the loyal but hierarchically organized collaboration between Koch, Gaffky, and Pfeiffer benefited all parties involved.

The relation between a *Doktorvater* and his former students was often more complicated, however. The correspondence between Wundt and Julius Merkel shows a somewhat awkward relationship.[44] Wundt occasionally criticized his former student so harshly that he was almost driven to despair: "Your criticism of my work, for which I would like to thank you sincerely, is […] so devastating that it could well take away all my courage to continue working in this way for even one more minute."[45] In his defence against Wundt's critique, Merkel repeatedly emphasized his loyal adherence to Wundt's school of thought: "After all, I have emphasized in the paper that I already sent you two years ago, that […] of all attempts to explain Weber's law, only the one originating from you comes into consideration." No matter how harshly Wundt criticized him, Merkel did not question his former teacher's accomplishments in any way. The letters of the other struggling former students of Wundt discussed in the third chapter, Friedrich Kiesow, Ludwig Lange, and Gottlob Theodor Lipps, did not contain such criticism either.

It would not be fair, however, to characterize Wundt as someone who pressured his former students into following his research programme. His most successful students all developed their own research interests and methods. Modern-day historians therefore agree that Wundt failed to establish his own school of thought. Edward Haupt argues that Georg Elias Müller's laboratory in Göttingen had taken over Wundt's lead in innovating experimental psychology in the 1890s.[46] Edwin Boring notes that Wundt's main influence may have been that "almost all the new schools have been founded as a protest against some one or other characteristic of Wundt's psychology."[47] Wolfgang Mack, finally, argues that Wundt's philosophical legacy did not fare any better than his psychological experimentalism; hardly anyone has bothered to engage with it.[48]

Wundt's reluctance to impose his own ideas on former students who were able to develop independent research programmes is illustrated by their widely diverging careers. Emil Kraepelin became one of Germany's foremost psychiatrists, Oswald Külpe founded the Würzburg school of psychology, Ernst Meumann's primary interest was pedagogy, and Hugo Münsterberg is mostly remembered today for his contributions to applied psychology.[49] Even though Wundt and Külpe disagreed about a lot of issues, the latter gladly remembered the former's encouragement to think for himself: "The more independent it is, the more it will delight me."[50] When Wundt ended his co-editorship of a journal managed by Meumann, he emphasized his appreciation of their friendship and stated that "each has to follow his own star."[51] He was even more explicit in a letter to Münsterberg: "I do not ask for gratitude [...]. If, however, someone wants to show his gratitude by his own choice, this can only happen in this way, by working solidly, diligently, and carefully, without caring about authorities or his career."[52] Of course, as I have set out in the introduction, this did not in any way mean that the relationship between Wundt and Münsterberg was unproblematic.[53]

The above examples show that an influential scholar like Wundt could demand loyalty of his students in at least two different ways. Those who were unable to make an independent career were asked to carry out research that fit his own programme. In exchange for their loyal efforts, Wundt provided them with a publishing platform and occasionally even with a temporary job. Those who were able to make an independent career were free to pursue their own interests, as long as they lived up to certain basic standards of good scholarship, such as working "solidly, diligently, and carefully." Wundt showed his loyalty to these independent students by helping them to find permanent academic employment. This explains why Richard Avenarius, Meumann, and Lipps were all hired to Wundt's former Chair of Inductive Philosophy in Zürich.[54] Wundt's promotion of his former students for a permanent university position was quite typical. De Goeje, for example, also lobbied the professoriate and the Mayor of Leiden to secure a professorial chair for his best student, Christiaan Snouck Hurgronje.[55] And even if Wernicke never studied with Behring, the latter's efforts to bring his friend and colleague to Marburg also fit this pattern.

A former student was expected to remain loyal to his *Doktorvater*. Of the protagonists of this study, Nöldeke was the only one to break with his teacher, Heinrich Ewald.[56] When Nöldeke started teaching in Göttingen alongside Ewald in 1860, their relationship quickly deteriorated. Looking back, he recounted that "Ewald was without even a trace of humour. Whoever had a little bit of independence, could not stay on good terms with him in the long run."[57] Nöldeke also felt uneasy about the way in which Ewald tended to cast himself in the role of a guardian of Christian faith and accuse his peers of all kinds of moral and religious shortcomings. He often criticized Heinrich Leberecht Fleischer, who mockingly commented to De Goeje that "in the end, we fully owe it only to [Ewald's] magnanimity and forbearance that we quasi-scholars still exist; if he would wish to destroy us, it would take him, like the JHWH of the Old Testament, just one breath and we would be gone!"[58] As soon as Nöldeke was appointed as a full professor, he wrote Ewald a letter in which he stated that—even though he would "never stop honouring [him] as a groundbreaking researcher and my foremost teacher"—he had to reserve for himself "the full freedom to express [his] honest scholarly convictions unhindered by personal considerations."[59] Ewald was—of course—furious, but De Goeje and Fleischer fully approved.

The Peer Group and the Reputation of the Discipline

Creating and maintaining group solidarity among a scattered group of peers is not easy. Regular face-to-face meetings facilitate such solidarity, but most scholars in late nineteenth-century Germany had more contact with their everyday colleagues than with their peers in faraway places. This is true even if an increasing number of scientific congresses encouraged scholarly sociability in the late nineteenth century.[60] Visits to international congresses, which have recently received considerable attention, were only minor features of the everyday working lives of most scholars.[61] Access to such congresses was often limited; participation required an invitation or a nomination by one's government. Therefore, only the most well-established scholars tended to attend. National and local congresses were more accessible. National scholarly societies, such as the *Deutsche Morgenländische Gesellschaft*, often organized annual meeting with a plenary assembly meeting, research presentations, and opportunities for socializing. Such meetings were only attended by a relatively small percentage of the society's membership, however. The list of the attendees of the meeting of the *Morgenländische Gesellschaft* in Trier 1879 contains thirty-seven names of which twenty-five were society members. The list of the society's members in that same year mentions more than 400 individuals as well as several institutional members.[62]

Congresses did not always instil feelings of solidarity. Behring complained that his colleagues disliked him and therefore refused to invite him to their congresses.[63] Wundt found little use for attending congresses and informed Külpe that he was "not a friend of congresses anyway, least of all of the international ones."[64] De Goeje's former student Snouck Hurgronje frequently complained about the lack of scholarly merit of the international congresses that he attended. He argued that the 1888 International Orientalist Congress in Stockholm was merely a celebration of its organizer, count Carlo Landberg: "[E]veryting, fun and seriousness, was aimed at putting one man [...] in a false magic light."[65] Almost twenty-five years later, Snouck complained that a congress in Athens had "more or less been swallowed by the celebration of the anniversary of the university."[66] So even though congresses could provide an opportunity for "banding together" and developing a shared "identity as professionals," they only made a limited contribution to the creation of a national—or even international—community of scholars.[67] In this light, the book reviews discussed in the fourth chapter become even more important.

The relative weakness of personal ties made it easy to assert one's own scholarly independence by critically assessing other scholars' work. Highly critical reviews were widely accepted. In the fourth chapter, we have encountered very personal accusations of dilettantism and carelessness, reproaches of dogmatism and mysticism, antisemitic dog whistling, and jokes ridiculing the reviewed work. What is more striking, however, is that most reviews were either neutral or favourable. One reason for this might be that most reviewed works deserved this benevolent judgement. Another reason is the fact that most editors and publishers were unwilling to print a high number of unfavourable reviews.[68] The number of positive assessments was even further increased by the common practice of writing reviews as a favour to friends. A final explanation for the many benevolent reviews can be found in the reviewers' feelings of loyalty toward the scholarly community to which both reviewer and reviewee belonged.

Nöldeke was very explicit about this reason for favourable reviewing in a letter to Ignaz Goldziher in which he praised him for the subtlety of his criticism in a book review: "Your mildness succeeds in hiding from laymen [...] that the work has no particular value, while you point that out sufficiently for the person, who is somewhat knowledgeable. I remember that Benfey [...] sometimes said that reviews of scholarly works should be such that they do not put the non-scholarly world at risk to develop a low opinion of the scholarship at which it is directed."[69] Snouck praises Goldziher for similar reasons in his obituary: "[He] never forgot the respectful piety, which he deemed mandatory for him to show to his seniors [and] the necessary indulgence of the weaknesses of his contemporaries." Sometimes, Snouck continued, the mildness of his reviews was "almost deceptive."[70]

These examples show that book reviews were considered to be potentially damaging to the reputation of the disciple. This potential risk, however, also provided an opportunity to express loyalty to the peer group by writing largely favourable reviews.

The most striking expressions of loyalty towards the peer group as a whole in book reviews can be observed in the different politeness strategies used by reviewers to ameliorate the face-threatening qualities of criticism and to emphasize the unity of purpose between reviewer, reviewee, and peer group. Through explicit expressions of solidarity, the underlining of their own lack of specialist knowledge, the acknowledgement of the usefulness of even imperfect scholarship, the praise of the ambition of the reviewee's aims, and even an occasional joke, reviewers emphasized that the whole disciplinary community shared a dedication to the same set of values. These values included diligence, clarity of exposition, and intellectual modesty. Every scholar who proved to be able to practice these virtues deserved the mild benevolence that would safeguard his reputation in the eyes of both his peers and a lay audience.

Institutions I: Scholarly Journals

Journals did not only provide a medium through which independence could be asserted and loyalty could be expressed. As institutional players in the academic landscape, journals invited a range of very specific assertions of independence and loyalty as well. Their editors, who tended to be scholars themselves, had a variety of reasons to establish or manage a journal. In order to achieve their goals, they had to balance the demands of a range of stakeholders that included publishers, contributors, and readership. Because different journal had different aims and different institutional backgrounds, the relationships between editors and each of these stakeholders could vary wildly. These variations resulted in different balances between independence from and loyalty towards each individual stakeholder.

The editors highlighted in the third chapter had very different relations to their publisher. Unlike the *Zeitschrift der Deutschen Morgenländischen Gesellschaft* and the *Literarisches Centralblatt für Deutschland*, the *Philosophische Studien* were the brainchild of one man: Wilhelm Wundt. His status as a best-selling author provided him with a strong bargaining position with his long-time publisher, Wilhelm Engelmann. This enabled Wundt to establish a journal in which he was completely free to offer an almost exclusive platform to his own students, friends, and other collaborators. He was therefore able to use the independence that his bargaining position vis-à-vis his publisher allowed him to be a loyal promotor of the work of his closest associates. Because his best-selling books already paid for his editorial endeavours, he did not have to worry about establishing a base of paying subscribers either.

Heinrich Leberecht Fleischer and August Fischer, the editors of the *Zeitschrift*, did not have to worry about paying subscribers either: their journal was institutionally embedded in the *Deutsche Morgenländische Gesellschaft* and all its paying members received the *Zeitschrift*. They could also operate quite independently from their publisher. There were close personal ties between the society and the Brockhaus publishing house, while the guaranteed readership of society members secured the financial viability of the endeavour. The *Zeitschrift*'s status as a society organ did, however, come with its very own editorial challenges. Fleischer and Fischer were supposed to manage the journal in such a way that all members felt free to share their insights in its pages while they also had to make sure that none of the society's members would take offense to the *Zeitschrift*'s contents. This aim could only be reached approximately by following three rules of thumb.

The first editorial rule of thumb was a requirement of basic politeness and abstention from *ad hominems*.[71] Offensive turns of phrase could be deleted from submitted papers and contributors were praised for their "spirit of moderation," "mutual respect," and "avoidance of all personal remarks and offensive insinuations."[72] The second rule of thumb stated that all members of the society had the right to publish a critical response to any article published in the *Zeitschrift*. Because of this rule, the editors were occasionally forced to publish responses that they thoroughly disliked. It also explains why the decision to reject a—admittedly rude—reply caused a lot of outrage.[73] The final rule of thumb stipulated that no debate should be allowed to linger on to prevent prolonged polemics. A contributor would get one opportunity to reply to a critical response and that would be the end of the discussion. Through this three-rule system, the editors of the *Zeitschrift* were able to balance the demands of independent criticism and loyal collegiality.

Friedrich Zarncke, the editor of the *Centralblatt* largely stuck to the same editorial rules of thumb, even if Gutschmid's antisemitic dog whistles show that he was not in the position to keep all potentially inflammatory remarks out of its pages.[74] Zarncke's greatest challenge, however, was to secure a base of paying subscribers. The constant struggle to attract such a base limited his editorial independence. He was forced to make sure that the *Centralblatt* reviewed more books than its short-lived competitor, the *Jenaer Literaturzeitung*, to compete in the niche market for scholarly review journals. He was also forced to consider his publisher's proposals to make the journal more attractive to a wider audience. He also had to make sure that the published reviews would not be too long to safe space for advertisements. These editorial constraints show that the moral economy that shaped the *Centralblatt*'s balance between loyal collegiality and critical independence cannot be understood in isolation from the demands of the economy of the marketplace.

Institutions II: Universities, Faculties, and Research Institutes

The most important institution in everyday academic life was, of course, the university and especially the faculty or institute where a researcher was employed. Worries about the functioning of one's faculty or research institute provided the perfect excuse to be critical of individual colleagues. This is particularly obvious in the example of the *Seminar für Orientalische Sprachen* (Institute for Oriental Languages) discussed in the preceding chapter.[75] Carl Arendt defended the harshness of his assessment of his colleagues by pointing at the interests of the institute: "Everywhere I have attempted to hold back my own *judgement*, I have, however, attempted to assemble with a certain completeness those facts that I know of, that seem suitable to contribute to the clarification of the not very pleasant overall atmosphere at the Institute."[76] Martin Hartmann likewise justified his scathing criticism by pointing at "the state of affairs [at the institute], which cannot but lead to severe damage and complications."[77] Considerations about the best interest of a faculty or institute also occurred frequently, however, at places that did not suffer from infighting and soured personal relations.

Often, the criticism of individuals in the light of the interest of institutions was not directed at the academic merit of the scholar in question. None of Althoff's informants, for example, expressed any doubts about Otto Heubner's qualities as a paediatrician.[78] Even his harshest critic, Eduard Külz, admitted that some of his works were indeed important and that his lectures were "not bad."[79] His reservations about Heubner were fully grounded in his conception of the University of Berlin as the elite teaching institution of the German *Reich*. This was a widely shared sentiment. In 1918, the historian Max Lenz, for example, emphasized that its university had been indispensable in turning Berlin into "a sanctuary and centre of German nature and art and all intellectual freedom."[80] In Külz's eyes, Heubner's Saxon accent and unpretentious demeanour did not suit the dignity of such a prestigious and metropolitan institution. His loyalty to this ideal outweighed his appreciation of Heubner's scholarly, educational, and managerial merits.

The discussions about the school of thought of a prospective new philosophy professor in Berlin in 1894 was also more about institutional interests than about the research—or even teaching—qualities of the candidates.[81] Wundt was never seriously considered for this position because he was not well-liked among Berlin's most prominent philosophers. The final choice was therefore between Carl Stumpf and Georg Elias Müller. All of Althoff's informants agreed that both men fitted the profile of a researcher with both psychological and experimental experience. Eventually, the choice fell on Stumpf, not because he was a better, more meticulous, more influential, or more innovative researcher, but because his additional interests in ethics and aesthetics suited the sensibilities and interests of the most influential members of the Berlin faculty.

In his recommendations to Althoff, Heinrich Fritsch even refused to take the interests of any of his individual peers into account. He was only concerned with the interests of the struggling medical faculty of the University of Breslau.[82] In the light of the low number of patients in the university clinic and the similarly low number of students, he successfully lobbied to have no successor of Friedrich Müller. His later recommendation of Otto Küstner for the Chair of Gynaecology was fully based on his assessment of the very specific needs of the Breslau Faculty of Medicine. Only a man of independent means, Fritzsch argued, could be expected to forego private practice, and pay his full attention to dealing with the demanding working conditions at his struggling faculty.

The above examples illustrate that the recommendations of Althoff's informants mostly discussed questions of fitting in. If the emphasis is on the fit between a faculty or institute on the one hand and individual scholars on the other, institutional interests tended to outweigh individual interests. After all, it is more difficult to reorganize a faculty or research institute to fit one scholar's specific strengths, weaknesses, and idiosyncrasies than to simply appoint someone who suits his prospective peers' preferences and self-image. The reports of Althoff's informants therefore primarily served as a way to express institutional loyalty. The Ministry of Education usually took these reports very seriously. Sometimes, however, Althoff's objectives would be in conflict with the wishes of the faculties. A well-organized faculty would occasionally be able to assert its independence vis-à-vis the ministry.[83] If the matter was important enough for Althoff, however, he was willing and able to use the state's full power. In these cases, it was almost impossible for a faculty to act on its proud assertions of independence.

Institutions III: Non-academic Institutions

Scholars did not only maintain relations with individual peers and academic institution but also with a variety of non-academic institutions. One obvious non-academic relation is between scholars and their publishers. The commercial aspirations of the Avenarius, Brill, and Engelmann publishing houses touched on the careers of Zarncke, De Goeje, and Wundt.[84] Loyalty and independence were central to the relation between author and publisher. Engelmann secured Wundt's loyalty by allowing him to publish a commercially uninteresting journal and De Goeje profited from the loyal support of Brill to carry out his long-running al-Ṭabarī' project. Zarncke continuously experienced that he could not assert full editorial independence: his publisher kept pushing him to adjust the *Centralblatt* to what he perceived to be the demands of the marketplace.

At times, scholars were expected to show loyalty not only to a peer group and easily identifiable institutions but also to more abstract entities like "society" or "the nation." At the outbreak of the First World War, many German intellectuals rushed to the defence of the German war objectives.[85] In the early days of the war, Wundt gave a lecture at one of Leipzig's biggest venues, the *Alberthalle* of the *Krystallpalast*, where thousands gathered to hear him chastise Germany's opponents and advocate the expansion of Germany's sphere of influence.[86] The lecture was published to great acclaim under the title *Über den wahrhaften Krieg* (On the True War).[87] Semitists joined the patriotic discussion about the war as well. Nöldeke's younger colleague and friend Carl Heinrich Becker, for example, published several booklets eulogizing the close relationship between Germany and its new ally, the Ottoman Empire.[88]

The outbreak of the First World War was an exceptional situation: for men like Wundt and Becker, it was relatively rare to find themselves at the centre of national attention. Bacteriologists like Koch and Behring, however, had to deal with the high expectations of a broad audience throughout their careers. After all, finding cures for deadly diseases is even more important to people suffering from them than to a researcher with primarily academic interests. Their breakthroughs were extensively covered by the news media, who often portrayed them as noble warriors, fighting for a brighter future without deadly diseases. They were even depicted as heroes in cartoons. Robert Koch was drawn as a brave horseman wielding his microscope to fight off the deadly snake of tuberculosis.[89] Behring was depicted as answering to the desperate prayers of a mother by wresting her child from the greedy arms of Death himself.[90]

The examples of the World War and the fight against deadly diseases are rather straightforward: it was clear what loyalties were expected. No German philosopher of Semitist would accidently argue against his country's war aims or against the merit of the German-Ottoman alliance. Likewise, no bacteriologist would distance himself from the widely acknowledged societal relevance of fighting deadly and contagious diseases. The expectations of different non-academic institutions could, however, come into conflict with each other. And—maybe even more alarming—loyalties towards non-academic actors and institutions could come into conflict with scholarly commitments. The career of Behring provides an illuminating example of how the cultivation of close relations to both state and industry could be used to gain a position of far-reaching independence within the academic community.

After the efficacy of Behring's diphtheria serum was established, it was produced and distributed by the *Höchster Farbwerke* from 1894 onward. The *Farbwerke* sought to diversify their output when the profitability of their main product, dyes, declined in the 1880s.[91] The collaboration with Behring was one of their most successful attempts at such diversification: in its first year, the serum generated a profit of 444,200 Mark and in its second year this increased to 764,800 Mark.[92] Because Behring's first five-year contract with the *Farbwerke* stipulated that he would get half of these profits, he received more than half a million Mark during the first two years of the commercial production of his serum.[93] The enormity of this figure is best understood in comparison to the lump sum that academic best-seller author Wundt received around the same time for his *Grundriss der Psychologie*: a comparatively modest 2,000 Mark.[94] The striking success of the diphtheria serum meant a big change for Behring's relations with the pharmaceutical industry, the Prussian government, and his bacteriological peers.

The efficacy and marketability of the serum—of course—endeared him to the management of the *Höchster Farbwerke*, which tried to recruit him as the firm's new director of research.[95] The serum's success also changed his relationship with Althoff for the better: the *Ministerialdirektor* was now convinced that Behring should be retained for the Prussian university system. At the same time, his relationship with other bacteriologists deteriorated. We have already seen that the financial success of the serum threatened the relationship with his closest friend and collaborator, Wernicke.[96] Behring also found that his colleagues occasionally refused to invite him to their conferences any longer.[97] The relationship that suffered most from Behring's sudden success and his subsequent attempts to achieve more independence was the one with Paul Ehrlich.

Ehrlich had made essential contributions to the development of the diphtheria serum in 1893.[98] Of course, he hoped to profit from the serum's success as well. However, August Laubenheimer, one of the directors of the *Farbwerke*, convinced him to refrain from any immediate monetary rewards and promised him the directorship of an independent organization for serum testing, instead.[99] Ehrlich could not resists the pressure of Laubenheimer and Behring, but soon came to regret this. Koch's Institute for Infectious Diseases was commissioned to test the serum and because Ehrlich was too close to the *Farbwerke*, he was not even considered to be part of the undertaking.[100] When he was finally put in charge of the newly established *Institut für Serumforschung und Serumprüfung* (Institute for Serum Research and Serum Testing) in 1896, Behring's attitude towards it bothered Ehrlich. Behring seemed to perceive of the Institute as an auxiliary organization in support of his own research rather than as an independent organization.[101]

The relationship between Behring and Ehrlich never recovered. Althoff was convinced that both men could play an important part in the development of new serums and tried to repair it. Ehrlich, however, was determined to keep Behring at arm's length. Collaborating with him, he argued, could only be expected from "a slavish character, but not from an independent researcher with the greatest thirst for freedom (such as I am, after all)."[102] Even Ehrlich's wife Hedwig wrote to Althoff to underline his reluctance to collaborate with Behring: "I know exactly that my husband can only work successfully when he [...] enjoys the full freedom of his creative work. [...] If, despite everything, he would decide to work with Behring, I want to beg you that his free status will be guaranteed and that he can terminate this association at any moment."[103]

While Behring's personal relations with his peers suffered, he cultivated his relationship with both the *Farbwerke*'s Laubenheimer and with Althoff. He used his leverage with Laubenheimer to secure a measure of independence from the state and fell back on his relationship with Althoff to guarantee his independence towards the *Farbwerke*. Behring was continuously on his guard not to become what he called a "slave of the industry."[104] This was the main reason he continued to pressure Althoff into finding a professorial chair for him: "How would I be able to guarantee my authoritative and independent position toward the *Höchster Farbwerken* when me and my institute are dependent on their goodwill. How would I be able [...] to follow up on the threat that I will immediately associate myself with another industrial enterprise?"[105] Althoff's successful struggle with the Marburg Faculty of Medicine on Behring's behalf shows that he was susceptible to this line of reasoning.

While he was negotiating with the University of Marburg, Althoff's proposed that Behring would immunize twelve horses as a first step to the establishment of a governmental serum production facility.[106] Behring, however, flat-out refused to comply: "[A]s long as a more profitable practical occupation has not been secured for me [...], I cannot decide [...] to give up the cooperation with the works so successfully carried out at the *Höchster Farbwerke*."[107] The threat to work exclusively for the *Farbwerke* was credible because by the late nineteenth century, collaborations between academics and industry had become quite respectable, not least because industrial enterprises were increasingly managed by university graduates.[108] Behring also hinted at leaving academia altogether in his negotiations with the *Farbwerke*. He mused about throwing off "all still existing chains" that prevented him from financially benefitting from his research and stated that if Althoff would find him a professorial chair his "hand would be even more tied than ever before." In his attempts to show his loyalty to the *Farbwerke*, he even suggested dedicating the book he was working on to Laubenheimer![109]

Behring's cultivation of his relations with Althoff and Laubenheimer allowed him to gain independence from the two conditions that had shaped his career before his invention of the diphtheria blood serum: his employments at Koch's Institute for Infectious Diseases and his status as a military doctor. The main downside of his military appointment and his employment under Koch was that he could not patent any of his inventions and innovations for commercial purposes. As he wrote to Laubenheimer, if he would remain a military doctor at the Berlin institute, he would "indeed have the fame of having accomplished a scientific achievement in the eyes of later generations" but he would, alas, "have to leave the fruits of his labour to the colleagues with business experience."[110]

The Dynamics and Complexity of Moral Economies

Each of the scholars discussed above maintained relations with a wide variety of interested parties. Friends, colleagues, students, (former) teachers, universities, faculties, research institutes, editors, professional societies, state authorities, and businesses competed for their loyal commitment. At the same time, they aspired to be independent researchers, teachers, and public intellectuals. The greatest challenge in maintaining independence was therefore to find a balance between the demands and expectations of all these different stakeholders. At the same time, however, it was exactly this myriad of ties that enabled individual scholars to define themselves as independent members of an academic community. Their independence consisted in choosing their loyalties and in strategically addressing different people and institutions to accomplish their career goals. The choices of the protagonists of the earlier chapters illustrate this observation.

Nöldeke's decision to distance himself from his *Doktorvater* Ewald, for example, was only announced after two conditions were met. Firstly, he had finally found a permanent job as a full professor in Kiel and, secondly, he had become part of the influential group of scholars that had taken shape around Fleischer. De Goeje's management of al-Ṭabarī's *Annals* provides another example of cleverly addressing various parties. If he would have relied on the support of his own university and closest colleagues, he would never have been able to finish his ambitious undertaking. With the loyal support of his publisher, government authorities, different professional societies, and a large international group of peers, he was able to finish the three-decade long project. Later in his career he would again collaborate with the same publisher, an international consortium of scholars, several professional societies, and government authorities to publish the *Encyclopedia of Islam*, the first edition of which would be published after his death under the supervision of his former students Christiaan Snouck Hurgronje and Theodoor Houtsma.[111]

Throughout his career, Wundt was able to obtain a level of independence through publishing. He could afford it to quit his first salaried but subordinate academic position, a time-consuming assistantship in Hermann von Helmholtz's Heidelberg laboratory, because he was able to support himself with his book revenues.[112] Later in his life was able to convince Engelmann to publish his financially unattractive *Philosophische Studien* because the sales numbers of his books were sufficiently high to make up for eventual losses. His student Münsterberg is another example of a psychologist who advanced his academic career by cultivating ties with people outside his peer group. Among other things, he collaborated with Althoff on the founding of the *Amerika Institut* and the *Internationale Wochenschrift*.[113] Münsterberg hoped—in vain—that his contributions to Althoff's projects would encourage him to support his professorial ambitions in a similar way as he supported Behring's ambitions.

The bacteriologists and hygienists at the Institute for Infectious Diseases had even more opportunities to address interested parties. The development of new drugs was followed with great interest by both the Ministry of Education and the Ministry of War. The commercial value of such drugs ensured the continuous interests of industrial manufacturers, such as the *Höchster Farbwerke*, as well. In addition, the search for new curative agents attracted wealthy private donors too. This allowed Paul Ehrlich to carry out independent research beyond his tasks at the Institute for Serum Research and Serum Testing. His investigations into a potential chemotherapeutic cure for cancer were largely made in the so-called *Georg-Speyer-Haus*, a research institute financed by the bequest of the banker Georg Speyer, after mediation by the industrialist and amateur historian of science Ludwig Darmstaedter.[114]

Even though involvement with non-academic parties could buy academic independence, it also carried risks. Especially the patenting and commercial exploitation of inventions could lead to the charge that one's work was inspired by a love for money rather than by a dedication to scholarship.[115] An overt love for money was not only considered to be in conflict with internationally shared ideals of scholarship but also with the self-image of the German bourgeoisie.[116] August Wilhelm Hoffmann—a main player in the emergence of the close relationship between academic chemists and the industry in the nineteenth century—articulated a widely agreed-upon consensus when he downplayed his industrial endeavours and emphasized his "pure fascination with the knowledge of truth and beauty, free from the pursuit of material benefit."[117] Behring's statements about not wanting to be a "slave of the industry" echoed this sentiment as well. What is more, his worries were justified by Ehrlich's repeated attempts to discredit him in Althoff's eyes by highlighting his financial interests.[118] Behring also had to defend himself against accusations of greed published in professional journals.[119] And, to add insult to injury, even the popular press occasionally made fun of his supposed love of money![120]

All in all, the case studies in this book show that the moral economy of scholarship that guided the relation between collegial loyalty and critical independence consisted of several different balancing acts. In the first place, scholars had to balance both virtues in their relationships with various individuals in academia, such as colleagues, students, supervisors, and collaborators. Secondly, they were supposed to be able to find a balance between both virtues in their relationship with the institutions at the heart of scholarship, such as universities, research institutes, scholarly journals, and professional organizations. Finally, they were supposed to deal with expectations that were not exclusively grounded in conceptions of scholarship, including such widely differing commitments as those to friendship, industry, the state, and the nation.

Of course, not all people and institutions fell clearly within only one of these categories. The editor of a scholarly journal could be seen both as a colleague and as the representative of his journal. A research institute could be seen as first and foremost an environment for research or as the materialization of the preferences of the state of other financiers. An official at the Ministry of Education could either be seen as an influential individual with whom one could cultivate a friendship or as an inescapable representative of the state. The fact that the definition of such relationship was rarely unambiguous underlines just how challenging it was to find an appropriate balance between the claims to loyalty of all the different parties that shaped the everyday working life in scholarly environments.

There was not one obvious formula to deal with these ambiguities. Scholars had to decide about loyalty and independence on a case-by-case level. Should a book review be treated as an opportunity to promote the work of a friend or as an opportunity to showcase your critical and independent thinking? Should a recommendation for a professorial chair be guided by a sense of loyalty towards your most valued peers, or by the interests of the institution advertising the vacancy, or even by your assessment of the interests of a scholarly or national community? To what extent should you be willing to allow state and industry actors to shape your research agenda? To what extent can loyalty to your peers only be shown by "by working solidly, diligently, and carefully, without caring about authorities or [your] career?"[121] To what extent does loyalty require you to share the financial benefits of your work with your collaborators? None of these questions could be answered on the basis of simple rules of thumb.

It should be emphasized, however, that loyalty and independence were not always at odds with each other. A close friendship could provide the mutual trust needed to engage in a detailed and thorough criticism of each other's work. Most book reviews combined words of praise with hedged criticisms. Journal editors developed rules of thumb that allowed for discussion while making sure that debates would not devolve into polemics. As more and more university researchers developed ties to different industries, the careers of men like Behring showed that it was indeed possible to contribute to the growth of scholarship from the cutting edge between state, industry, and academia.

Even though the protagonists of this book worked in very different disciplines, the continuous struggle to balance loyal collegiality and independent criticism was central to the everyday working lives of all of them. The moral economies that shaped their careers and working days were more complex than those described by Kohler and McCray. Kohler has drawn attention to a moral ethos of communality that suggests a high level of appreciation for loyalty towards peers and academic institutions. This valuation of loyalty was the result a state of abundance that encouraged the sharing of data and *Drosophila* cultures. This wealth of data and cultures, then, made it easy to engage in independent research within a larger loyal community of scholarship. McCray, on the other hand, describes a scholarly economy of scarcity: research institutes were pitted against each other in a competition for scarce resources. This environment encourages assertions of independence even if the need to share materials forced researchers to get along. My broader study of the moral economy of scholarship across disciplines in late nineteenth-century Germany, however, adds to the insights of Kohler and McCray by highlighting the dynamic character and the complexity of these moral economies. Scholars in all disciplines were continuously involved in a struggle to balance their loyalty to and independence from their scholarly peers, academic institutions, and a wide range of non-academic stakeholders.

Notes

1. See Chap. 1, 7–8.
2. Clifford Geertz, *The Interpretation of Cultures: Selected Essays* (New York, NY: Basic Books, 1973), 7.
3. Robert E. Kohler, *Lords of the Fly: Drosophila Genetics and the Experimental Life* (Chicago, IL: University of Chicago Press, 1994), 92–93.
4. Ibid., 123–124.
5. Ibid., 167.
6. Patrick W. McCray, "Large Telescopes and the Moral Economy of Recent Astronomy," *Social Studies of Science* 30, no. 5 (2000): 685–711.
7. Lorraine Daston, "The Moral Economy of Science," *Osiris* 10 (1995): 4.
8. See Chap. 2, 47.
9. Michael Jan de Goeje to Theodor Nöldeke, 15 December 1863, Universiteitsbibliotheek Leiden (hereafter UBL): BPL 2389.
10. Theodor Nöldeke to Michael Jan de Goeje, 14 October 1907, UBL: BPL 2389.
11. Michael Jan de Goeje to Theodor Nöldeke, 15 December 1863, UBL: BPL 2389.
12. Theodor Nöldeke to Michael Jan de Goeje, 16 December 1863, UBL: BPL 2389.
13. Michael Jan de Goeje to Theodor Nöldeke, 20 December 1863, UBL: BPL 2389.
14. Examples of De Goeje praising Dozy's book in private or public include: Michael Jan de Goeje to Theodor Nöldeke, 3 February 1864, UBL: BPL 2389; Michael Jan de Goeje, "Een stap vooruit," *De Gids* 28, no. 2 (1864): 297–312. The broader discussion that broke out after the publication of Dozy's book is discussed in: Herman Paul, "Virtue language in nineteenth-century orientalism: a case study in historical epistemology," *Modern Intellectual History* 14, no. 3 (2017): 689–715.
15. Theodor Nöldeke to Michael Jan de Goeje, 12 June 1864, UBL: BPL 2389.
16. Wilhelm Wundt, *Reden und Aufsätze* (Leipzig: Alfred Kröner, 1914), 315. Wundt would be involved with the Fechner estate for a long time. In 1894, seven years after Fechner's death, he approached Gottlob Friedrich Lipps to edit one of Fechner's unfinished manuscripts. The project would take several years: Wilhelm Wundt to Gottlob Friedrich Lipps, 15 February 1894, Universitätsarchiv Leipzig (hereafter UAL): NA Wundt/III/501–600/569/515–518; Wilhelm Wundt to Gottlob Friedrich Lipps, 1 December 1898, UAL: NA Wundt/III/501–600/576/547–550.
17. See Chap. 2, 43.

18. Gustav Theodor Fechner to Wilhelm Wundt, 18 June 1879, UAL: NA Wundt/III/1701–1723/1712/1b/17–44
19. On Fechner's presence at Friedrich Zöllner's sessions with Slade, see: Matthias Schneid, *Der neuere Spiritismus philosophisch geprüft* (Eichstätt: August Hornik, 1880), 28. On Fechner's declining eyesight, see: Wundt, *Reden und Aufsätze*, 313.
20. Gustav Theodor Fechner to Wilhelm Wundt, 28 June 1879, UAL: NA Wundt/III/1701–1723/1712/1b/17–44.
21. Gustav Theodor Fechner to Wilhelm Wundt, 25 June 1879, NA Wundt/III/1701–1723/1712/1c/45–48.
22. See Chap. 2, 43–44.
23. Michael Jan de Goeje to Theodor Nöldeke, 13 August 1864, UBL: BPL 2389.
24. Emil Behring to Erich Wernicke, 22 November 1891, Behring Nachlass digital (herafter BNd): EvB/B 1/177.
25. Emil Behring to Erich Wernicke, 15 February 1892, BNd: EvB/B 1/186.
26. Wernicke wrote: "*Die Freundschaft ist kein leere Wahn!*" This is a variation on "*Und die Treue, sie ist doch kein leerer Wahn*" in Schiller poem *Die Bürgschaft*, a ballad in praise of friendship and loyalty. Wernicke entry in the guest book is dated 22 April 1897 and can be found at: BNd: EvB/L 266.
27. Emil Behring to Erich Wernicke, 11 March 1892, BNd: EvB/B 1/190; Emil Behring to Erich Wernicke, 16 October 1894, BNd, EvB/B 1/220.
28. Emil Behring to Erich Wernicke, 25 February 1897, BNd, EvB/B 1/244
29. Emil Behring to Erich Wernicke, 2 January 1899, BNd: EvB/B 1/248.
30. Emil Behring to Erich Wernicke, 30 December 1899, BNd: EvB/B 1/259.
31. Emil Behring to Erich Wernicke, 12 December 1908, BNd: EvB/B 1/273.
32. BNd: EvB/L 266.
33. David Cahan, "The institutional revolution in German physics, 1865–1914," *Historical Studies in the Physical Sciences* 15, no. 2 (1985): 5.
34. Alexander Busch, "The Vicissitudes of the *Privatdozent*: Breakdown and Adaption in the Recruitment of the German University Teacher," *Minerva* 1, no. 3 (1963): 328.
35. Jeffrey A. Johnson, "Hierarchy and Creativity in Chemistry, 1871–1914," in *Science in Germany, The Intersection of Institutional and Intellectual Issues*, Osiris, second series, 5, ed. Kathryn M. Olesko (1989): 215.
36. On Gaffky, see: Jost Benedum, "Georg Gaffky (1850–1918) und die Gründung des Gießener Lehrstuhls für Hygiene im Jahre 1888," *Gießener Universitätsblätter* 221 (1989): 91–94.
37. On Pfeiffer, see Ernst Th. Rietschel and Jean-Marc Cavaillon, "Richard Pfeiffer and Alexandre Besredka: creators of the concept of endotoxin and anti-endotoxin," *Microbes and Infection* 5 (2003): 1407–1414.

38. Georg Gaffky to Robert Koch, 2 October 1892, Robert Koch Institut (hereafter RKI), as/b1/003.
39. Ibid.
40. This is shown by the signature of Karl Köhler, the director of the *Kaiserliche Gesundheitsamt*, on several letters that Gaffky addressed to Koch: Georg Gaffky to Robert Koch, 2 October 1892, RKI, as/b1/003; Georg Gaffky to Robert Koch, 12 October 1892, RKI, as/b1/005.
41. Karl Köhler to Robert Koch, 9 October 1892, RKI, as/b1/004; Georg Gaffky to Robert Koch, 12 October 1892, RKI, as/b1/005.
42. For a modern-day assessment of Pettenkofer's insights, see Wolfgang Gerhard Locher, "Max von Pettenkofer (1818–1901) as a Pioneer of Modern Hygiene and Preventive Medicine," *Environmental Health and Preventive Medicine* 12 (2007): 238–245; Alfredo Morabia, "Epidemiologic Interactions, Complexity, and the Lonesome Death of Max von Pettenkofer," *American Journal of Epidemiology* 166, no. 11 (2007): 1233–1238.
43. Robert Koch, "Die Cholera in Deutschland während des Winters 1892 bis 1893," *Zeitschrift für Hygiene und Infektionskrankheiten* 15 (1893): 117; Richard Pfeiffer to Robert Koch, 5 February 1893, RKI, as/bi/007.
44. See Chap. 3, 70.
45. Julius Merkel to Wilhelm Wundt, 8 October 1887, UAL: NA Wundt/ III/601–700/674/339–346.
46. Edward J. Haupt, "Laboratories for Experimental Psychology: Göttingen's Ascendancy over Leipzig in the 1890s," in *Wilhelm Wundt in History: The Making of a Scientific Psychology*, ed. Robert W. Rieber and David K. Robinson (New York, NY: Springer Science + Business Media, 2001), 205–250.
47. Edwin G. Boring, *A History of Experimental Psychology*, second edition (Englewood Cliffs, NJ: Prentice-Hall, 1950), 343.
48. Wolfgang Mack, "Wundts programmatische Erbe," in *Wilhelm Wundts anderes Erbe: Ein Missverständniss löst sich auf*, ed. Gerd Jüttemann (Göttingen: Vandenhoeck & Ruprecht, 2006), 241.
49. On Kraepelin, see: Holger Steinberg (ed.), *Der Briefwechsel zwischen Wilhelm Wundt und Emil Kraepelin: Zeugnis einer jahrzehntelangen Freundschaft* (Bern: Hans Huber, 2002). On Külpe, see: Horst Gundlach, "Oswald Külpe und die Würzburger Schule," in *Hundert Jahre Institut für Psychologie und Würzburger Schule der Denkpsychologie*, ed. Wilhelm Janke and Wolfgang Schneider (Göttingen: Hogrefe Verlag für Psychologie, 1999), 107–124; David Lindenfeld, "Oswald Külpe and the Würzburg School," *Journal of the History of the Behavioral Sciences* 14 (1978): 132–141. On Meumann, see: Paul Probst, *Bibliographie Ernst Meumann: Mit einer Einleitung zur Biographie* (Herzberg: Traugott

Bautz, 1991). On Münsterberg, see: Phyllis Keller, *States of belonging: German-American intellectuals and the First World War* (Cambridge, MA: Harvard University Press, 1979), 21–30; Frank J. Landy, "Hugo Münsterberg: Victim or Visionary?" *Journal of Applied Psychology* 77, no. 6 (1992): 787–802.

50. Oswald Külpe to Wilhelm Wundt, 22 September 1895, UAL: NA Wundt/III/301–400/390/283–290.
51. Wilhelm Wundt to Ernst Meumann, 20 December 1904, UAL: NA Wundt/III/701–800/723a/149–150.
52. Wilhelm Wundt to Hugo Münsterberg, 12 November 1890, UAL: UAL, NA Wundt/III/701–800/765/427–438.
53. See Chap. 1, 1–3.
54. Münsterberg was almost hired in Zürich as well. When his appointment was delayed, he spoilt his chances with an angry and impatient letter to the Zürich Faculty of Philosophy. Theodor Vetter to Wilhelm Wundt, 20 March 1897, UAL: NA Wundt/III/1501–1600/1519/65–68.
55. Michael Jan de Goeje to Theodor Nöldeke, 4 November 1888, UBL: BPL 2389. The Mayor of Leiden was one of the custodians of Leiden University. In the end, Snouck would not accept the professorship in Malay that De Goeje had arranged for him. He joined the administration of the Dutch East Indies and would only return to academia in 1906 to succeed De Goeje.
56. I discuss in more detail in Christiaan Engberts, "Gossiping about the Buddha of Göttingen: Heinrich Ewald as an Unscholarly Persona," *History of Humanities* 1, no. 2 (2016): 371–385.
57. Theodor Nöldeke to Ignaz Goldziher, 22 February 1904, Magyar Tudományos Akadémia Könyvtár Információs Központ (Library and Information Centre of the Hungarian Academy of Sciences (hereafter MTAK): GIL/32/01/160.
58. Heinrich Leberecht Fleischer to Michael Jan de Goeje, 11 March 1865, UBL: BPL 2389.
59. Theodor Nöldeke to Heinrich Ewald, 4 October 1869, quoted in: Enno Littmann, "Theodor Nöldeke," in *Nachrichten von der Gesellschaft der Wissenschaften zu Göttingen: Geschäftliche Mitteilungen aus dem Berichtsjahr 1930/31* (Berlin: Weidmannsche Buchhandlung, 1931), 56.
60. Wolf Feuerhah and Pascale Rabault-Feuerhahn, "Présentation: la science à l'échelle internationale," *Revue germanique internationale* 12 (2010): 5.
61. A relevant recent study on international congresses is Debra Adrienne Everett-Lane, International Scientific Congresses, 1873–1913: Community and Conflict in the Pursuit of Knowledge (PhD diss., Columbia University, 2004).

62. "Nachrichten über Angelegenheiten der Deutschen Morgenländischen Gesellschaft," *Zeitschrift der Deutschen Morgenländischen Gesellschaft* 34 (1880): I–LXV.
63. Ferdinand Hueppe to Emil von Behring, 13 February 1907, Geheimes Staatsarchiv Preußischer Kulturbesitz (hereafter GStA PK), VI. HA, Nl Althoff, no. 668; Emil von Behring to Friedrich Althoff, 15 Feburary 1906, GStA PK, VI. HA, Nl Althoff, no. 668.
64. Wilhelm Wundt to Oswald Külpe, 1 January 1909, UAL, NA Wundt/III/401-500/417/89–92.
65. Christiaan Snouck Hurgronje to Theodor Nöldeke, 4 May 1890, quoted in: Pieter S. van Koningsveld (ed.), *Orientalism and Islam: the letters of C. Snouck Hurgronje to Th. Nöldeke from the Tübingen University Library* (Leiden: Documentatiebureau Islam-Christendom, 1985), 19. Snouck was not the only visitor to be disappointed with the Stockholm Congress, see: Pascale Rabault-Feuerhahn, "'La science la robe au vent' Le Congrès International des Orientalistes et la disciplinarisation des études Orientales," in: *La Disciplinarisation de la linguistique. Histoire et épistémologie, Historie épistémologie langage*, Dossier en ligne 5 (2012): 1–16.
66. Christiaan Snouck Hurgronje to Theodor Nöldeke, 7 May 1912, quoted in Koningveld, *Orientalism and Islam*, 95.
67. Everett-Lane, International Scientific Congresses, 222–223.
68. See Chap. 3, 85.
69. Theodor Nöldeke to Ignaz Goldziher, 2 January 1910, MTAK: GIL/32/01/235. Theodor Benfey taught Sankrit in Göttingen when Nöldeke studied and taught there.
70. Christiaan Snouck Hurgronje, "Ignaz Goldziher. 22 juni 1850–13 november 1921," *De Gids* 85 (1921): 494.
71. This rule of thumb is also recognized by Raf de Bont: Raf de Bont, "'Writing in letters of blood': Manners in scientific dispute in nineteenth-century Britain and the German lands," *History of Science* 51 (2013): 309–335. He also notes that this rule was not always successfully enforced.
72. See Chap. 3, 76 and Heinrich Leberecht Fleischer to Theodor Nöldeke, 21 July 1875, Universitätsbibliothek Tübingen (hereafter UBT): Md 782 A 68.
73. See Chap. 3, 79.
74. See Chap. 3, 87–88.
75. See Chap. 5, 143–146.
76. Professor Arendt über die Zustände am Seminar für Orientalische Sprachen, 4 VII 1889, GStA PK, VI. HA, Nl Althoff, no. 156. Arendt's emphasis.
77. Martin Hartmann to 'Sehr geehrter Herr Kreisschulinspektor,' 26 September 1889, GStA PK, VI. HA, Nl Althoff, no. 156.

78. See Chap. 5, 151–152.
79. Eduard Külz to Friedrich Althoff, 13 November 1893, GStA PK, VI. HA, Nl Althoff, no. 112.
80. Max Lenz, *Geschichte der königlichen Friedrich-Wilhelms-Universität zu Berlin*, zweiter Band, zweite Hälfte: auf dem Wege zur deutschen Einheit im neuen Reich (Halle an der Saale: Verlag der Buchhandlung des Waisenhauses, 1918), 357.
81. See Chap. 5, 160–164.
82. See Chap. 5, 146–149.
83. Hartmut Leppin, "Eduard Nordens Berufung nach Greifswald: Handlungsspielräume im 'System Althoff'," *Philologus* 142, no. 1 (1998): 162–172.
84. See Chap. 3, 89–90.
85. Wolfgang J. Mommsen, "Einleitung: Die deutschen kulturellen Eliten im Ersten Weltkrieg," in *Kultur und Krieg: Die Rolle der Intellektuellen, Künstler und Schriftsteller im Ersten Weltkrieg*, ed. Wolfgang J. Mommsen (Munich: Oldenbourg, 1996), 1–15.
86. Steffen Bruendel, *Volksgemeinschaft oder Volksstaat: Die 'Ideen von 1914' und die Neuordnung Deutschlands im Ersten Weltkrieg* (Berlin: Akademie, 2003), 36–37.
87. Wilhelm Wundt, *Über den wahrhaften Krieg* (Leipzig: Alfred Kröner, 1914).
88. Carl Heinrich Becker, *Deutschland und der Islam* (Stuttgart and Berlin: Deutsche Verlags-Anstalt, 1914); Carl Heinrich Becker, *Deutsch-türkische Interessengemeinschaft* (Bonn: Friedrich Cohen, 1914). I discuss these booklets in more detail in Christiaan Engberts, "Orientalists at War: Personae and Partiality at the Outbreak of the First World War," in *Scholarly Personae in the History of Orientalism*, ed. Christiaan Engberts and Herman Paul (Brill: Leiden, 2019): 172–192.
89. B. Lee Ligon, "Robert Koch: Nobel Laureate and Controversial Figure in Tuberculin Research," *Seminars in Pediatric Infectious Diseases* 13, no. 4 (2002): 296.
90. Behring der Retter der Kinder, BNd: EvB/S 27.
91. Timothy Lenoir, "A Magic Bullet: Research for Profit and the Growth of Knowledge in Germany Around 1900," *Minerva* 26, no. 1 (1988): 70.
92. According to a note that reads "*Diphtherie-Heilserum*": BNd, EvB/B 196/115. I take this to be a list of the annual profits of the serum because a report by *Farbwerke* director Laubenheimer states that the profit for 1894 was 444,200, which corresponds to the 1894 number on the note: BNd, EvB/B 196/7.

93. Vertrag zwischen Emil von Behring und den Farbwerken Vormals Meister, Lucius & Brüning betreffend 'Gewinnung von Diphtherie-Heilserum", § IV, BNd, EvB/B 196/2/4.
94. Emanuel Reinicke to Wilhelm Wundt, 14 October 1895, UAL, NA Wundt/III/1601–1700/1683/1/1–4.
95. See Chap. 5, 153.
96. See this chapter, 181.
97. Ferdinand Hueppe to Emil von Behring, 13 February 1906; Emil von Behring to Friedrich Althoff, 15 February 1906, GStA PK, VI. HA, Nl Althoff, No. 668.
98. See Chap. 2, 39.
99. Axel C. Hüntelmann, *Paul Ehrlich: Leben, Forschung, Ökonomien, Netzwerke* (Göttingen: Wallstein Verlag, 2011), 104.
100. Ibid., 106.
101. Paul Ehrlich to Friedrich Althoff, 12 September 1903, GStA PK, VI. HA, Nl Althoff, No. 709.
102. Ibid.
103. Hedwig Ehrlich to Friedrich Althoff, 13 Sepbember 1903, GStA PK, VI. HA, Nl Althoff, No. 709.
104. Emil von Behring to Die Direktion der Farbwerke Höchst/Man, 27 July 1903, BNd: EvB/B 196/78.
105. Emil Behring to Friedrich Althoff, 8 February 1895, GStA PK, VI. HA, Nl Althoff, No. 325.
106. Friedrich Althoff to Emil Behring, 15 November 1894, GStA PK, VI. HA, Nl Althoff, No. 326.
107. Emil Behring to Friedrich Althoff, 17 November 1894, GStA PK, VI. HA, Nl Althoff, No. 325.
108. Peter Borscheid, *Naturwissenschaft, Staat und Industrie in Baden (1848–1914)* (Stuttgart: Ernst Klett, 1976), 115, 127.
109. August Laubenheimer to Emil Behring, 5 April 1894, BNd: EvB/B 196/10. As far as I know, Behring never dedicated a book to Laubenheimer in the end.
110. EmilBehringotAugustLaubenheimer,30March1894,BNd:EvB/B196/9.
111. For a detailed overview of the realization of the *Encyclopedia of Islam*, see Peri Bearman, *A history of the Encyclopaedia of Islam* (Atlanta, GA: Lockwood Press, 2018) and Daniel van der Zande, Martinus Th. Houtsma: 1851–1943: Een bijdrage aan de geschiedenis van de Oriëntalistiek in Nederland en Europa (PhD diss., Universiteit Utrecht, 1999), 225–260.
112. Wolfgang G. Bringmann, Norma J. Bringmann and William D.G. Balance. "Wilhelm Maximilian Wundt 1832–1874: The Formative Years." In

Wundt Studies: A Centennial Collection, ed. Wolfgang G. Bringmann and Ryan D. Tweney (Toronto: C.J. Hogrefe, 1980), 25–27.

113. On Münsterberg's involvement with the *Amerika Institut*, see Christian H. Freitag, Die Entwicklung der Amerikastudien in Berlin bis 1945 (PhD diss., Freie Universität Berlin, 1977), 39–49. For his involvement with the Wochenschrift, see: Minutes of the conference on the *Internationale Monatsschrift* by Wilhelm Lexis, 2 January 1907, GStA PK, VI. HA, Nl Althoff, No. 295. The initial plan to publish a *Monatsschrift* (monthly journal) was eventually replaced by the plan to publish a *Wochenschrift* (weekly journal) instead.
114. Hüntelmann, *Paul Ehrlich*, 280–284.
115. For a discussion about the awkwardness surrounding patents, see Paul Lucier, "Court and controversy: patenting science in the nineteenth century," *British Journal for the History of Science* 29 (1996), 141. For a discussion of the relation between wealth and scholarship, see Léjon Saarloos, The scholarly self under threat: Language of vice in British scholarship (1870–1910), PhD diss., Universiteit Leiden, 2021), 45–52.
116. Christina von Hodenberg, "Der Fluch des Geldsacks. Der Aufstieg des Industriellen als Herausforderung bürgerliche Werte," in *Der bürgerliche Wertehimmel: Innenansichten des 19. Jahrhunderts*, ed. Manfred Hettling and Stefan Ludwig Hoffmann (Göttingen: Vandenhoeck & Ruprecht, 2000), 79–104.
117. Quoted in Borscheid, *Naturwissenschaft, Staat und Industrie*, 125.
118. Paul Ehrlich to Friedrich Althoff, [no date, presumably January 1900], BNd: EvB/B 196/1; Paul Ehrlich to Friedrich Althoff, 17 February 1906, GStA PK, VI. HA, Nl Althoff, No. 668.
119. Emil Behring, "In eigener Sache," *Deutsche Medizinische Wochenschrift* 37, 15 September (1898), 595.
120. Althoff kept a copy of the short-lived Berlin satirical magazine *Das Narrenschiff*, that printed a fictional report of Behring convincing an unwitting reporter to pay for a newly developed serum against the "hardening of the heart." GStA PK, VI. HA, Nl Althoff, No. 358.
121. See this chapter, note 52.

CHAPTER 7

Epilogue: Loyalty and Independence Today

The Past and the Present

My primary interest in this study has been to sketch the way in which loyal collegiality and critical independence shaped the careers and everyday work experiences of nineteenth-century scholars. To illustrate the manifold different entanglements of these virtues, I have made a transdisciplinary comparison between orientalists, experimental psychologists, and bacteriologists. Because members of these disciplinary communities engaged in different research practices and had to deal with different expectations of a variety of interested parties, their experiences could be compared at the level of the moral economies that shaped individual scholars' attempts to establish a delicate balance between the demands of loyalty and independence.

The focus on a comparative investigation of moral economies of the past did not leave much room for reflection on role of loyalty and independence in modern-day academic life. To anyone working in academia today, however, the continued significance of loyal collegiality and critical independence might seem almost self-evident. After all, most researchers still aim for intellectual independence and original research as well as a supportive and collegial work environment. At the same time, it cannot be taken for granted that the ways of establishing equilibriums of nineteenth-century researchers have not changed throughout an eventful century.

© The Author(s), under exclusive license to Springer Nature Switzerland AG 2022

C. Engberts, *Scholarly Virtues in Nineteenth-Century Sciences and Humanities*, https://doi.org/10.1007/978-3-030-84566-7_7

Twentieth-century scholarship has gone through many changes and it would be surprising if this would not have left any mark on contemporary moral economies.

Despite all these changes, however, it would also be surprising if the moral demands placed on modern academics would not show any trace of the demands placed on men like Emil Behring, Robert Koch, Theodor Nöldeke, and Wilhelm Wundt. In this epilogue, I will first sketch some of the major twentieth-century changes in the ways in which scholarship is organized. I will subsequently draw attention to the way in which older reflections on loyalty and independence are still relevant in contemporary academia. To be sure, I do not aim to give a comprehensive analysis of these virtues' standing among twenty-first-century researchers. I only have the more modest aim to invite the reader to reflect on the continued significance of collegial loyalty and critical independence, two virtues of scholarship that I have so far discussed as elements of past scholarship but that many might recognize as strikingly familiar features of their everyday working lives.

A Century of Change?

The outbreak of the First World War exposed the fragility of nineteenth-century moral economies of scholarship. In an upsurge of "scientific nationalism," German scholars with long and successful academic careers reassessed commitments that they had long taken for granted.[1] Of course, the late nineteenth-century growth of German science and the construction of a German nation after the unification of 1871 had been entwined processes: scholars knowingly contributed to "creating the shared high culture that would make for national identity."[2] In addition, even staunch internationalists acknowledged that the international character of science was grounded in "competition between national scientific champions."[3] The outbreak of the First World War, however, forced scholars to reassess their priorities and many concluded that—for the time being—loyalty to the nation outweighed all other virtues of scholarship. Carl Heinrich Becker acknowledged as much in a letter to his Dutch friend and colleague Christiaan Snouck Hurgronje: "For us Germans, even our scholarly ideals fade into the background as insignificant in this great time, now that our fatherland is struggling for its existence."[4] Becker's and Wundt's lectures and publications in the early years of the World War also attest to their sudden feelings of patriotism.[5]

The outbreak of the First World War was only one of the many events that would reshape the scholarly landscape of the twentieth century. The Second World War and the Cold War also left their marks on research and teaching in and outside of universities, as did the processes of decolonization and globalization. In the light of the themes discussed in the preceding chapters, however, the transformation of the practices and ideals of scholarship that touch on the funding of scholarly research, its organizational structure, the relation between academia and industry, the degree of international collaboration, and the culture of scholarly publishing are particularly relevant.

The funding of research has changed considerably during the last century. Throughout the nineteenth century, German academic research was largely funded by the government, even if Behring's association with the *Höchster Farbwerke* and Paul Ehrlich's involvement with the *Georg-Speyer-Haus* are some early examples of the entanglement of academia and the private sector.[6] In the twentieth century, private funding by wealthy foundations like the John D. Rockefeller Foundation and the Andrew Carnegie Foundation became increasingly important.[7] The highly competitive selection procedures developed by these foundations aimed to ensure that only the most promising research proposals would be funded. The twentieth-century successes of American science persuaded European governments to establish new agencies for the promotion of scientific research.[8] These organizations were often modelled after the private foundations and adopted similar competitive selection processes. The operations of Germany's Kaiser Wilhelm Institutes and the *Notgemeinschaft der Deutschen Wissenschaft*, for example, were modelled on both the Rockefeller and the Carnegie foundations.[9]

These new funding arrangements went hand in hand with another change in the organizational structure of research, the emergence of what is commonly referred to as Big Science. The grand collaborative efforts that are designated as such, unsurprisingly, have at least some historical precedents. Some argue that the roots of Big Science—both in the humanities and the STEM fields—can be traced back to the nineteenth century. The group of scholars working on Theodor Mommsen's *Corpus Inscriptiorum Latinarum* and the physical and chemical laboratories of the time are appealing examples of these early roots.[10] Michael Jan de Goeje's al-Ṭabarī edition is another example of early large-scale scholarly collaboration.[11] The awareness of this new development in the organization of scholarly work is also reflected in Adolf Harnack's critical commentary on the emerging "big enterprise of science."[12]

Most modern-day commentators, however, associate the term Big Science with the "big organizations, big machines, and big politics" that became dominant during the Cold War.[13] Alvin Weinberg, who coined the term, explicitly described twentieth-century Big Science as "much more complicated" than earlier "Little Science" and as requiring "extremely elaborate equipment and staffs of large teams of professionals."[14] Derek de Solla Price also underlines discontinuity with his observation that "if any sufficiently large segment of science is measured in any reasonable way, the normal mode or growth is exponential."[15] The development of Big Science is intertwined with the development of new funding mechanisms because expensive large-scale projects often depend on the wealth of the newly founded private and public research foundations. The collaborative ambitions and increasing costs of twentieth-century Big Science also contributed to the emergence of international funding bodies, such as the European Research Council (ERC).[16]

The growth of Big Science has been accompanied by an increasingly intricate entanglement of academic research and industry. This growing interconnectedness manifests itself in different ways. Twentieth-century researchers and their academic employers have, for example, demonstrated a growing interest in the commercial potential of their work. A recent study even starts from the assumption that it "is clear that universities need to become more entrepreneurial, changing their strategies, their structures and their practices, changing their culture and helping students and faculty members to develop their entrepreneurial mindsets and entrepreneurial actions."[17] At the same time, a growing number of university-educated researchers has found employment in the industrial sector. Steven Shapin notes that, as early as the 1950s, more than half of all US scientists and research engineers was working in the industry.[18] Even though the career of Behring has shown that collaborations between industry and university existed in nineteenth-century Germany, the modern-day convergence of entrepreneurship and academia has become so inescapable that the argument can be made that the "sale of science is a relatively new phenomenon, and it follows the modern ways of business, rather than the ancient ways of science."[19]

Another development related to the emergence of Big Science and the accompanying entanglement of academia and industry is the increasingly international character of scholarship. Of course, the nineteenth century had its fair share of international congresses.[20] Of course, De Goeje's al-Ṭabarī consortium provides another example of a nineteenth-century international collaboration. After the Second World War, however, science has become more international than ever before. This is perhaps best illustrated by the emergence of English as the only dominant language of scholarship. Michael Gordin even refers to contemporary scientists as "the most resolutely monoglot international community the world has ever seen."[21] This is in stark contrast with earlier linguistic practices. The proceedings of the international orientalist congress in Hamburg in 1902, for example, were quite typical in accepting papers in the three major languages of science: English, French, and German.[22] The Italian attendees submitted their contributions in their own language, which had been a major language of science until the early nineteenth century.[23] The contribution of the Albanian poet, philologist, and former Jesuit Ndre Mjeda was even published in the scholarly *lingua franca* of an earlier era: Latin![24]

Foundation funding, Big Science, academic entrepreneurship, and the internationalization of science have heavily impacted the scholarly community. Men like Behring, Koch, Nöldeke, and Wundt were personally acquainted with most other influential researchers in their disciplines in their country. The modern-day international scholarly community, however, is too big to facilitate such close personal relationships between all—or even most—prominent researchers in a discipline. Among other things, this has created a more impersonal academic community in which a new publishing culture has emerged. The largely informal nineteenth-century evaluation of manuscripts submitted to journals has—quite recently—been replaced by the double-blind editorial peer review process which today is seen as "the lynchpin about which the whole business of Science is pivoted."[25] The example provided by the recent emergence of the peer review process suggests that personal relations have at least to some extent been replaced by impersonal structures. This is a development that also touches on the potential differences between past and current claims to loyalty and assertions of independence.

Loyalty and Independence in the Twenty-First Century

So, what impact have these changes in the organization of publishing, funding and collaboration had on the delicate balances between scholarly loyalty and independence? In the light of all these developments, it is tempting to simply conclude that the old moral economy simply must have been superseded. The argument could be made that the intimacy of private correspondence has been replaced by the anonymity of double-blind peer review, that the evaluative practices of Big Science do not require the commitment to the creation and maintenance of a community of peers that was promoted by Nöldeke's and Wundt's book reviews, that the pursuit of funding from different types of foundations cannot be compared in any meaningful way to the professorial appointment procedures of the Althoff system, or that the attachment to ostensible selfless contributions to the growth of knowledge is insurmountably far removed from an increasing willingness to monetize research findings.

The persistent interest in the virtues and vices of scholarship throughout the twentieth century, however, suggests a continued relevance of reflections on the moral economy of science. The appointment policies of German universities and research institutes after the Second World War provide a first telling example of the pervasive and ambiguous way in which a virtue like loyalty shaped the twentieth-century academic environment. Mutual loyalty among researchers who thrived in Nazi Germany ensured that scholars who could not clear themselves from the suspicion of moral compromise were nevertheless able to continue their careers after the Second World War.[26] Reflections on the supposed moral failure of scholars and the damaging impact this could have on the perception of scholarly trustworthiness also resonate with a wide audience that is not concerned with such obvious moral atrocities as a Nazi past. In the early 1980s, for example, the journalists William Broad and Nicholas Wade hit a nerve with their study of fraud cases at prestigious US research institutes.[27] Their work instigated a lengthy debate about modern-day research ethics.[28] More recently, Horace Freeland Judson followed up on their work and discussed whether questions of research ethics can be meaningfully framed in terms of individual guilt and innocence.[29] Examples of individual researchers struggling to balance virtue and vice clearly continue to speak to the imagination.

The entanglement of academia and industry also receives some harsh criticism. Daniel Greenberg worries about the way in which "contemporary science is embedded in, and financed by, a society that worships money and profits and celebrates personal wealth" because this might not go together with the scientist's "sacred obligation" to seek for truth.[30] Steven Shapin is less alarmist. He defines the modern-day scientific entrepreneur as "one who is both a qualified scientist and, like all commercial entrepreneurs, a risk taker" with "one foot in the making of knowledge and the other in the making of artifacts, services, and, ultimately, money."[31] The twofold character of the modern-day research scientist, however, does not pose a threat to reflections on the moral aspects of scholarship. On the contrary, I would argue that these ambiguities are at the root of contemporary moral dilemmas that are similar to the moral considerations of the past mostly because they are best understood as questions of scholarly virtue. Shapin's central contention therefore is that "personal virtue, familiarity, and charisma feature in such characteristically later modern configurations as the industrial research laboratory and the entrepreneurial network."[32]

In a similar vein, I would like to argue that this study's reflections on the moral economy of scholarship are relevant to modern-day academia in at least three ways. In the first place, it draws attention to the continued significance of personal interaction in scholarship. The genre conventions of today's academic publishing tend to obscure this. Frederic Holmes describes the modern research article as a "story reduced to the elements deemed essential to its outcome, [pared] of all details of procedure and background" that leaves little room for the acknowledgment of personal indebtedness.[33] And even though recent developments, such as the publication of negative results and the sharing of data sets, add some transparency to the intricacies of scholarly research, the importance of personal interaction often remains obscure.[34] However, footnotes and acknowledgement sections in which individual contributors are thanked are very common. This suggests that informal support still matters greatly to the production and evaluation of knowledge today.

Secondly, even both of the virtues discussed in this study still speak to a lot of scholars. It can even be argued that loyal collegiality has gained a renewed importance in collaborative efforts inspired by the ideal of Big Science. In an environment in which teamwork has become increasingly important, this virtue seems essential. In this vein, the Nobel Prize laureate J. Michael Bishop has gratefully drawn attention to his experience that "collegiality is the great privilege of science, to be a part of that huge international community. It's probably the most cohesive and enlightened international community that exists."[35] The ideal of critical independence is also as alive as ever. The rationale behind double-blinded review processes, after all, is that the resulting anonymity guarantees the reviewers' critically independent attitude.

Thirdly, and finally, this study's focus on some of the virtues that shaped the moral economy of nineteenth-century German scholarship makes it easier to recognize the importance of new virtues in the age of entrepreneurial scholarship and Big Science. Shapin draws attention to some of these new virtues. Directors of industrial research centres, for instance, are expected "to be unusually persuasive," to be able to build up relationships "of trust with corporate headquarters," and to "actively manage the moral regimes of their laboratories."[36] In addition, they are expected to be able "to accept the reality of compromising about research agenda's," to know how to "interact with dissimilar people," and learn to accept that a company might demand unquestioning, absolute loyalty within a hierarchical environment.[37] Finally, they are also supposed to possess the virtues of "flexibility" and "adaptability" and become true "risk taker[s]."[38] All these new expectations and demands are explicitly framed in an older language of virtue that invites reflective comparison.

The virtues of the nineteenth century—collegial loyalty and critical independence—are still relevant to the everyday work environment and self image of modern day scholars. The task to find an appropriate balance between these virtues, however, has become even more challenging to the extent that today's scholars also must weigh in new virtues, like persuasiveness in a corporate environment, flexibility, and the ability and willingness to deal with the sort of risks posed by the market economy. Balancing old and new virtues is obviously very demanding. Some contemporary scholars, therefore, choose to assert their independence from the claims of the modern neoliberal university and emphasize ideals of solidarity and loyalty instead. They are advocates of what has become known as "slow scholarship" and encourage an ethics of care for oneself and others, "deep reflexive thought, engaged research, joy in writing and working with concepts and ideas driven by our passions."[39] The highlighting of these values is not fundamentally different from the virtue language of late nineteenth-century German scholars.[40] Like their precursors, modern-day scholars have to find some sort of balance between virtues and expectations. Like their precursors, they also often find that this hard-won balance remains subject to change. The ensuing continuous effort to re-evaluate and recalibrate the balance between loyalty, independence, and virtues that have become important more recently, lies at the heart of the twenty-first-century moral economy of scholarship.

Notes

1. Elisabeth Crawford, *Nationalism and internationalism in science, 1880–1939: Four studies of the Nobel Population* (Cambridge: Cambridge University Press, 1992), 28–46.
2. Ibid., 33.
3. Paul Forman, "Scientific Internationalism and the Weimar Physicists: The Ideology and Its Manipulation in Germany after World War I," *Isis* 64, no. 2 (1973), 153.
4. Carl Heinrich Becker to Christiaan Snouck Hurgronje, 17 September 1914, Universiteitsbibliotheek Leiden (hereafter UBL): Or. 8952 A: 149. For more details, see Christiaan Engberts, "Orientalists at War: Personae and Partiality at the Outbreak of the First World War,." in *Scholarly Personae in the History of Orientalism*, ed. Christiaan Engberts and Herman Paul (Leiden: Brill, 2019).
5. See Chap. 6, 191.
6. See Chap. 6, 192–195.
7. Pieter Huistra and Kaat Wils, "Fit to Travel: The Exchange Programme of the Belgian American Educational Foundation: An Institutional Perspective of Scientific *Persona Formation (1920–1940)," Low Countries Historical Review* 131, no. 4 (2016), 112–134; Robert Kohler, *Partners in Science: Foundations and Natural Scientists 1900–1945* (Chicago, IL: University of Chicago Press, 1991), 3.
8. In 1950 the National Science Foundation was founded in the United Kingdom. In the same year the *Nederlandse Organisatie voor Wetenschappelijk Onderzoek* (NWO, Netherlands Organization for Scientific Research) and the National Science Foundation) was established. The *Deutsche Forschungsgemeinschaft* (DFG, German Research Foundation) was founded one year later.
9. Jeffrey Allen Johnson, "The Academic-Industrial Symbiosis in German Chemical Research," in *The German Chemical Industry in the Twentieth Century*, ed. John E. Lesch (Springer Science + Business Media: Dordrecht, 2000), 22; Ulrich Marsch, *Notgemeinschaft der Deutschen Wissenschaft: Gründung und frühe Geschichte, 1920–1925* (Frankfurt am Main: Peter Lang, 1994), 23; Paul Weindling, "The Rockefeller Foundation and German Biomedical Sciences, 1920–40: from Educational Philanthropy to International Science Policy," in *Science, Politics and the Public Good*, ed. Nicolaas A Rupke (Basingstoke: Macmillan Press, 1988), 121.
10. Lorraine Daston, *Before the Two Cultures: Big Science and Big Humanities in the Nineteenth Century* (Jerusalem: Israel Academy of Sciences and Humanities, 2015), 9–10; Mary Jo Nye, *Before Big Science: The Pursuit of*

Modern Chemistry and Physics 1800–1940 (Cambridge, MA: Harvard University Press, 1996), xvi.

11. I discuss the al-Ṭabarī edition as an early example of Big Science in more detail in: Christiaan Engberts, "Een taalkundige als internationaal ondernemer: M.J. de Goeje en de Annalen van al-Tabari (1872–1901)," *De Moderne Tijd* 4, no. 1/2 (2020): 36–54.
12. Adolf Harnack, "Vom Großbetrieb der Wissenschaft," *Preußischer Jahrbücher* 119 (1905): 195.
13. Olof Hallonsten, *Big Science Transformed: Science Politics and Organization in Europe and the United States* (New York, NY: Palgrave Macmillan, 2016), 17.
14. Alvin M. Weinberg, *Reflections on Big Science* (Cambridge, MA: M.I.T. Press, 1967), 39.
15. Derek J. De Solla Price, *Little Science, Big Science* (New York, NY: Columbia University Press, 1963), 4–5.
16. Barbara Hoenig, *Europe's New Scientific Elite: Social Mechanisms of Science in the European Research Area* (London: Routledge, 2017), 111–112.
17. Alain Fayolle and Dana T. Redford, "Introduction: towards more entrepreneurial universities—myth or reality?" in *Handbook on the Entrepreneurial University*, ed. Alain Fayolle and Dana T. Redford (Cheltenham: Edward Elgar, 2014), 1.
18. Steven Shapin, *The Scientific Life: A Moral History of a Late Modern Vocation* (Chicago, IL: Chicago University Press, 2008), 110.
19. Daniel S. Greenberg, *Science for Sale: The Perils, Rewards, and Delusions of Campus Capitalism* (Chicago, IL: University of Chicago Press, 2007), 2.
20. Elisabeth Crawford, *Nationalism and internationalism in science, 1880–1939: Four studies of the Nobel Population* (Cambridge: Cambridge University Press, 1992), 38–39.
21. Michael D. Gordin, *Scientific Babel: The Language of Science from the Fall of Latin to the Rise of English* (Chicago, IL: University of Chicago Press, 2015), 2.
22. *Verhandlungen des XIII. Internationalen Orientalisten-Kongresses. Hamburg September 1902* (Leiden: Brill, 1902).
23. Gordin, *Scientific Babel*, 10.
24. Andreas Miedia, "De pronunciatione palatalium in diversis albanicae linguae dialectis," in *Verhandlungen des XIII. Internationalen Orientalisten-Kongresses. Hamburg September 1902* (Leiden: Brill, 1902), 14–15. See also: George W. Gawrych, *The Crescent and the Eagle: Ottoman Rule, Islam and the Albanians, 1874–1913* (Londen: I.B. Tauris, 2006), 89.
25. On the relatively recent development of the peer review process, see: Alex Csiszar, *The Scientific Journal: Authorship and the Politics of Knowledge in the Nineteenth Century* (Chicago, IL: University of Chicago Press, 2018),

Chapter 3. One author even argues that editorial peer review was not yet ubiquitous until the nineteen-seventies, see: Melinda Baldwin, "Scientific Autonomy, Public Accountability, and the Rise of 'Peer Review' in the Cold War United States," *Isis* 109, no. 3 (2018): 543.

26. Carola Sachse, ""Whitewash Culture": How the Kaiser Wilhelm/Max Planck Society Dealt with the Nazi Past," in *The Kaiser Wilhelm Society under National Socialism*, ed. Susanne Heim, Carola Sachse and Mark Walker (Cambridge: Cambridge University Press, 2009), 373–399.
27. William Broad and Nicholas Wade, *Betrayers of the Truth: Fraud and Deceit in the Halls of Science* (New York, NY: Touchstone, 1982).
28. Michael Davis, *Ethics and the University* (London: Routledge, 1999), 49–52.
29. Horace Freeland Judson, *The Great Betrayal: Fraud in Science* (Orlando, FL: Harcourt, 2004), 8.
30. Greenberg, *Science for Sale*, 5.
31. Shapin, *The Scientific Life*, 210.
32. Ibid., 5.
33. Frederic L. Holmes, "Argument and Narrative in Scientific Writing," in *The Literary Structure of Scientific Argument: Historical Studies*, ed. Peter Dear (Philadelphia, PA: University of Pennsylvania Press, 1991), 180–181.
34. On the publication of negative results, see Christian Pfeffer and Bjorn R. Olsen, "Editorial: Journal of Negative Results in Biomedicine," *Journal of Negative Results in Medicine* 1, no. 2 (2002): 1–2. On data sharing, see: Christoph Schickhardt, Hosley Nelson and Eva C. Winkler, "Researchers' Duty to Share Pre-publication Data: From the Prima Facie Duty to Practice," in *The Ethics of Biomedical Big Data*, ed. Brent Daniel Mittelstadt and Luciano Floridi (Cham: Springer, 2016), 309–338.
35. J. Michael Bishop quoted in István Hargittai and Magdolna Hargittai, "J. Michael Bishop," in *Candid Science VI: More Conversations with Famous Scientists*, ed. István Hargittai and Magdolna Hargittai (London: Imperial College Press, 2006), 199.
36. Shapin, *The Scientific Life*, 162.
37. Ibid., 234–235.
38. Ibid., 210, 258–261.
39. Alison Mountz et al., "For Slow Scholarship: A Feminist Politics of Resistance through Collective Action in the Neoliberal University," *ACME* 14 (2015): 1253.
40. Herman Paul, *De deugden van een wetenschapper: Karakter en toewijding in de geesteswetenschappen, 1850–1940* (Amsterdam: Amsterdam University Press, 2018), 192.

Index[1]

[1] Note: Page numbers followed by 'n' refer to notes.

© The Author(s), under exclusive license to Springer Nature Switzerland AG 2022

C. Engberts, *Scholarly Virtues in Nineteenth-Century Sciences and Humanities*, https://doi.org/10.1007/978-3-030-84566-7

The manufacturer's authorised representative in the EU is Springer Nature Customer Service Centre GmbH, Europaplatz 3, 69115 Heidelberg, Germany. If you have any concerns regarding our products, please contact ProductSafety@springernature.com

Printed and bound by CPI Group (UK) Ltd, Croydon, CR0 4YY
30/09/2026
02225499-0004